# AN UPDATED CLASSIFICATION OF THE RECENT CRUSTACEA

BY
JOEL W. MARTIN
AND
GEORGE E. DAVIS



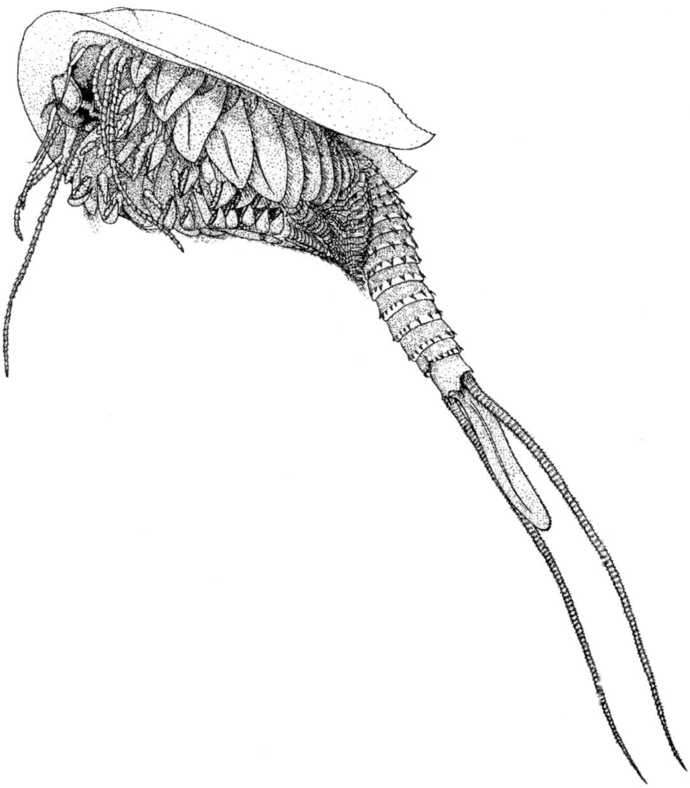

NO. 39
SCIENCE SERIES
NATURAL HISTORY MUSEUM
OF LOS ANGELES COUNTY

Natural History Museum of Los Angeles County
Los Angeles, California 90007
ISSN 1-891276-27-1
Published on 14 December 2001
Printed in the United States of America

# PREFACE

For anyone with interests in a group of organisms as large and diverse as the Crustacea, it is difficult to grasp the enormity of the entire taxon at one time. Those who work on crustaceans usually specialize in only one small corner of the field. Even though I am sometimes considered a specialist on crabs, the truth is I can profess some special knowledge about only a relatively few species in one or two families, with forays into other groups of crabs and other crustaceans. Crabs are but a small picture of the overall diversity of the Crustacea. They represent only one infraorder [Brachyura] within one order [Decapoda] within one superorder [Eucarida] within one subclass [Eumalacostraca] within one class [Malacostraca] of the six currently recognized classes of the Crustacea (as depicted herein). I am certain that this situation is similar for all other crustacean systematists, with the result that there are no living specialists who can truly claim to have an in-depth understanding of the Crustacea as a whole.

This volume is an attempt to provide the reader, whether a seasoned systematist or a beginning student, with a glimpse into the enormous variety of extant crustaceans. The sheer number of categories that humans have constructed to contain and order this group is some indication of the incredible amount of morphological diversity they exhibit. But this is only a small part of the overall picture. Even if one were to grasp the full range of taxonomic diversity as presented in this classification, such knowledge would shed no light on the actual biology of these fascinating animals: their behavior, feeding, locomotion, reproduction; their relationships to other organisms; their adaptations to the environment; and other facets of their existence that fall under the heading of biodiversity.

By producing this volume we are attempting to update an existing classification, produced by Tom Bowman and Larry Abele (1982), in order to arrange and update the Crustacea collection of the Natural History Museum of Los Angeles County. This enormous and diverse collection contains an estimated four to five million specimens, making it the second largest collection of Crustacea in the Americas. While undertaking this task, it occurred to us that others might benefit from our efforts, and that perhaps a general update on the number and arrangement of the living crustacean families, along with an explanation of the systematic and classificatory changes suggested during the last two decades, might be a welcome addition to the literature. I hope this volume is seen as nothing more than the briefest of introductions into an understanding of crustaceans and that it might lead to further work not only on the relationships among crustaceans but also toward understanding the overall picture of crustacean biodiversity and natural history.

Joel W. Martin
June 2001
Los Angeles, California

# ACKNOWLEDGMENTS

We sincerely thank the many carcinologists to whom we sent earlier versions of the classification (all of whom are listed in Appendix II). Although not all of these persons responded to our queries (we had a response rate of approximately 60% to the first mailing and approximately 70% to the second) and some saw only later versions, we felt it appropriate to list all persons from whom comments were solicited. Drs. Rodney Feldmann and Geoffrey Boxshall, in addition to commenting on sections of the classification, served as external referees for the entire manuscript, and to both we are extremely grateful. We mourn the loss of Erik Dahl in January 1999, of Mihai Băcescu in August 1999, of Arthur Humes and Austin Williams in October 1999, of Gary Brusca and Ray Manning in January 2000, of Théodore Monod in November 2000, and of Denton Belk in April 2001, during the compilation of this classification. Their absence is keenly felt by all carcinologists. Deserving of special recognition are David K. Camp for supplying much needed information and literature for a wide variety of taxa; Anne C. Cohen for literature on ostracodes and maxillopods and for enlightening discussions of that group's presumed monophyly; William Newman and Mark Grygier, both of whom provided literature and enlightening comments on maxillopods; Mark Grygier for additional comments on interpretation of ICZN recommendations; Trisha Spears and Cheryl Morrison for providing unpublished molecular sequence or gene rearrangement data for the decapods; Geoffrey Fryer for his always direct comments concerning the branchiopods; Gary Poore for information and literature on several peracarid and decapod groups and for his detailed review of our penultimate draft; Robert Hessler for providing needed literature and for his insightful suggestions; and Lipke Holthuis for suggesting corrections to several taxonomic authorities and dates in our earlier versions. Obviously, not all of the suggestions we received were incorporated, in part because some suggested changes contradicted others and in part because some suggested changes would have involved major rearrangements for which we deemed the evidence insufficient or incomplete. Inclusion of crustacean-related web sites as an appendix was the idea of Keith Crandall. We thank Todd Zimmerman, Regina Wetzer, Todd Haney, and Sandra Trautwein in our Los Angeles crustacean laboratory for suggestions and assistance at various points; Regina Wetzer in particular was instrumental in assembling Appendix III.

We thank the Natural History Museum of Los Angeles County, and especially John Heyning, John Harris, and the members of the Scientific Publications Committee, for support and for assistance with readying the manuscript for publication. We also thank the National Science Foundation for partial support via grants DEB 9020088, DEB 9320397, and DEB 9727188 to J. W. Martin; NSF Biotic Surveys and Inventories grant DEB 9972100 to T. L. Zimmerman and J. W. Martin; and NSF PEET grant DEB 9978193 to J. W. Martin and D. K. Jacobs. Finally, we sincerely thank Sue, Alex, and Paul Martin and Ruthe Davis for their kind encouragement and understanding.

# CONTENTS

# An Updated Classification of the Recent Crustacea

*By* Joel W. Martin[1] and George E. Davis[1]

**ABSTRACT.** An updated classification of the Crustacea down to the level of family is provided. The classification is based loosely on that given by Bowman and Abele (1982) and includes all new families and higher level taxa described since that time. In addition, in several crustacean groupings, new arrangements and assignments have been incorporated, based usually on phylogenetic information that has accrued or that has become more widely accepted since 1982. Among the more salient changes, some of which are more controversial than others, are the recognition of the former phylum Pentastomida as a group of maxillopod crustaceans based on additional spermatological and molecular evidence, the inclusion of the parasitic Tantulocarida also among the maxillopods, the treatment of the Branchiopoda as the most primitive extant group of crustaceans, and the recognition of Guinot's (1977, 1978) division of the higher (eubrachyuran) crabs into two "grades" based primarily on placement of the genital aperture. The revised classification includes 849 extant families in 42 orders and 6 classes; this is an increase of nearly 200 families since the Bowman and Abele classification. More than 90 specialists in the field were consulted and asked to contribute to the update. Some workers are not in agreement with our final arrangement. In particular, there are questions or dissenting opinions over our choice of which taxa to recognize, which authorities and dates to credit for various taxa, and especially over the arrangements among and/or within the higher taxa. As an aid to future workers in crustacean classification and phylogeny, comments and dissenting opinions of some of these workers are appended to highlight areas of uncertainty or controversy. Also appended are a list of the specialists who were given the opportunity to respond (Appendix II) and a list of printed and World Wide Web resources that contain information on crustaceans (Appendix III). The new classification is in part a result of one such site, the Crustacean Biodiversity Survey (formerly found at URL http://www.nhm.org/cbs/, now temporarily off-line).

## GENERAL INTRODUCTION

No group of plants or animals on the planet exhibits the range of morphological diversity seen among the extant Crustacea. This morphological diversity, or disparity in the paleontological jargon, is what makes the study of crustaceans so exciting. Yet it is also what makes deciphering the phylogeny of the group and ordering them into some sort of coherent classification so difficult. Because of the great age of the group, extending back at least as far as the early Cambrian and almost certainly beyond that, there has been ample time for endless experimentation with form and function. The result of these many millions of years of evolution is quite dazzling. The current estimate of the number of described species is approximately 52,000 (Land, 1996; Monod and Laubier, 1996). This estimate is surely on the low side, as a recent estimate of the number of living species of ostracodes alone is 10,000 to 15,000 (K. Martens, pers. comm., and discussions on the electronic ostracode listserver OSTRACON@LISTSERV.UH.EDU) and Kensley (1998) has estimated more than 54,000 for the reef-associated peracarids. Among the Metazoa, the estimate of 52,000 species places crustaceans fourth, behind insects, molluscs, and chelicerates, in terms of overall species diversity. But morphological diversity (disparity) is higher in the Crustacea than in any other taxon on Earth. There are probably few other groups of animals (squids come to mind because of *Architeuthis*) in which the difference in maximum size of adults can be a factor of 1,000. The known size of crabs now ranges from a maximum leg span of approximately 4 m in the giant Japanese spider crab *Macrocheira kaempferi* and a maximum carapace width of 46 cm in the giant Tasmanian crab *Pseudocarcinus gigas* (as cited in Schmitt, 1965) to a minimum of 1.5 mm across the carapace for a mature ovigerous female pinnotherid, *Nannotheres moorei,* the smallest known spe-

[1] Natural History Museum of Los Angeles County, Research and Collections, Department of Invertebrate Zoology, 900 Exposition Boulevard, Los Angeles, California 90007

Email: jmartin@nhm.org and gdavis@nhm.org

cies of crab (Manning and Felder, 1996). An ovigerous hermit crab (probably genus *Pygmaeopagurus*) with a shield length of only 0.76 mm taken from dredge samples in the Seychelles (McLaughlin and Hogarth, 1998) might hold the record for decapods, and of course much smaller crustaceans exist. Tantulocarids, recently discovered parasites found on other deep-sea crustaceans, are so small that they are sometimes found attached to the aesthetascs of the antennule of copepods; the total body length of *Stygotantulus stocki* is only 94 µm "from tip of rostrum to end of caudal rami" (Boxshall and Huys, 1989a:127). In terms of biomass, that of the Antarctic krill *Euphausia superba* has been estimated at 500 million tons at any given time, probably surpassing the biomass of any other group of metazoans (reviewed by Nicol and Endo, 1999). In terms of sheer numbers, the crustacean nauplius has been called "the most abundant type of multicellular animal on earth" (Fryer, 1987d). Crustaceans have been found in virtually every imaginable habitat (see Monod and Laubier, 1996), have been mistaken for molluscs, worms, and other distantly related animals, and continue to defy our attempts to force them into convenient taxonomic groupings. Indeed, there is still considerable debate over whether the group is monophyletic (see below).

Not surprisingly, the history of crustacean classification is a long and convoluted one. A summary of that history is well beyond the scope of this paper, and the reader is referred to the following publications as some of many possible starting points: Schram (1986); Fryer (1987a, c); Dahl and Strömberg (1992); Spears and Abele (1997); Rice (1980); Schram and Hof (1998); Monod and Forest (1996); and papers in the edited volumes *The Biology of Crustacea* (1982–1985; D. E. Bliss, editor-in-Chief) (especially volume 1); *Crustacean Issues* (F. R. Schram, general editor); *Arthropod Fossils and Phylogeny* (G. D. Edgecombe, editor); *Traité de Zoologie* (P.-P. Grassé, series editor; J. Forest, crustacean volumes editor); and the *Treatise of Invertebrate Paleontology* (R. C. Moore, editor) (a revision of this last work is currently underway). Despite the long history of studies on Crustacea, in many ways, we are just beginning our journey. New and significant finds continue to delight and surprise the student of the Crustacea. In the last two decades, the newly discovered taxa Remipedia, Tantulocarida, and Mictacea, as well as beautifully preserved fossils from the "Orsten" fauna of Sweden, are some of the more obvious examples. Another striking example of how little we know about crustaceans is the relatively recent discovery of an entirely new phylum of animal life, the Cycliophora (Funch and Kristensen, 1995; Winnepenninckx et al., 1998), found living on the mouthparts of the Norway lobster *Nephrops norvegicus,* a species of commercial importance that is encountered often in European restaurants.

The 1982 classification of the Recent Crustacea by T. E. Bowman and L. G. Abele, in turn based to a large extent on that of Moore and McCormick (1969), was a benchmark compilation that has been of tremendous use to students of the Crustacea. In that classification, the extant crustaceans were divided among 6 classes, 13 subclasses, 38 orders, and 652 families. Although it was recognized by Bowman and Abele and other workers in the field, even at the time of publication, that the classification was intended to be little more than a stopgap measure, it has continued to be employed in many major treatments of crustaceans (e.g., Barnes and Harrison, 1992; Young, 1998) and has widely influenced the study of crustaceans since its appearance. Subsequent to the appearance of the Bowman and Abele (1982) classification, a large number of new families and even some higher level taxa have been described. Indeed, our current list includes 849 families, an increase of 197 families over the Bowman and Abele (1982) classification. Thus, an argument could be made that an updated classification is warranted on the basis of the increased number of new families alone. A more compelling reason is that several major treatises have appeared that offer substantially different arrangements of those taxa and that many exciting areas of phylogenetic research and improved methodology have contributed significantly to our understanding of the relationships within the Crustacea and of the Crustacea to other arthropod groups.

While attempting to arrange the collections at the Natural History Museum of Los Angeles County, the second largest collection of crustaceans in the United States, we decided to update the Bowman and Abele (1982) classification by simply inserting the taxa described since that time. This proved to be a more difficult task than we originally envisioned. In part this was because the number of new taxa was larger than we first thought. And, in part, it was because there have been so many suggestions for new arrangements and groupings of crustacean assemblages, and we wanted to reflect some of the recent thinking in crustacean phylogeny in the arrangement of our museum's collection. At about the same time, we announced a World Wide Web product (http://www.nhm.org/cbs/) called the Crustacean Biodiversity Survey (Martin, 1996). The Survey was designed to allow workers from anywhere in the world to add information at a variety of levels to a database on crustacean biodiversity. The currently proposed classification is one result of that survey.

Lines have to be drawn at certain times in order to attain some level of completion. We received the suggestion from several workers to take the classification down to the level of subfamily; one worker even suggested we include a list of all known genera for each family. Others suggested that we provide a clear diagnosis and/or characters that distinguish each taxon or at least each major clade. Although these additions would undoubtedly be extremely helpful, for what we hope are obvious reasons, we did not want to attempt it. We are also aware that there are a number of works in progress that will have a bearing on our understanding of the classification of Crustacea (future volumes of the *Traité de Zoologie* [J. Forest, editor] and the ongoing revision of the Crustacea sections of the *Treatise on Invertebrate Paleontology* [edited by R. L. Kaesler, University of Kansas] are examples of works we have not yet seen). However, the field is moving rapidly, and we felt that there was more merit to publishing what we have than in waiting for additional analyses and publications to appear. We are also aware of the relatively recent suggestions to replace Linnaean hierarchical taxonomy and classification with a more phylogenetically based system. A brief review by Milius (1999, *Science News*, vol. 156: 268) outlines the controversy as presented at the International Botanical Congress meetings in St. Louis (see also de Queiroz and Gauthier, 1994; Hibbett and Donoghue, 1998; Cantino et al., 1999; Cantino, 2000; Nixon and Carpenter, 2000; Meier and Richter, 1992; and the web site for the PhyloCode at www.ohiou.edu/phylocode/). Some authors have even advocated doing away with species names as a supposedly logical consequence of using phylogenetic taxonomy (e.g., Pleijel and Rouse, 2000). However, we have retained a more classical approach for now.

## METHODS

To arrive at the present classification, we began by incorporating all of the changes or rearrangements of which we were aware. Mostly, because of our own taxonomic interests and the strengths of the Crustacea collection of the Natural History Museum of Los Angeles County, this meant the changes or updates within the Decapoda and Branchiopoda. In addition, we scanned the following journals from 1982 until the present: *Crustaceana, Journal of Crustacean Biology, Proceedings of the Biological Society of Washington, Smithsonian Contributions in Zoology, Contributions in Science of the Natural History Museum of Los Angeles County, Researches on Crustacea* (now *Crustacean Research*), and *Journal of Natural History.* Knowing that these journals would not provide a complete account of the many changes and additions suggested since 1982, we then endeavored to solicit the input of a large number of crustacean systematists from around the world. Any measure of completeness is due to the considerable help and input given by these workers (Appendix II). At the same time, we accept the responsibility and inevitable criticism that any such undertaking generates, as final decisions were made by us.

After incorporating comments received from the first mailing of the updated classification, we again sent the classification back to the same carcinologists and also to several other workers whose names had been suggested to us. Finally, in a third mailing, we asked those same workers (again, with some new names added to the list) to send us additional corrections and also their comments, supportive or otherwise, concerning the resulting classification, with the promise that we would try to publish these comments verbatim as Appendix I. In this way, we hope to point out areas of disagreement and existing controversies in the "current" classification such that future workers will know that what is presented here as a classification is merely a suggested starting point and that there is considerable room for improvement.

Not all workers responded. Some responded only to the first mailing, others only to the second or third. And of course not all persons listed in Appendix II received all three of the mailings. It is important to note that the listing of a name in Appendix II does not necessarily imply agreement with the new classification, regardless of whether a dissenting opinion has been offered. We also received a large number of positive comments and letters of encouragement.

The present classification will not be accepted by all current workers and is sure to be considered obsolete almost immediately. Yet we have found the Bowman and Abele (1982) classification to be of such help, in everything from organizing our museum collections to searching for taxa with which we are unfamiliar, that we hoped to provide a similar and updated tool that would be of at least some usefulness for students of the Crustacea.

As concerns the authorship of this paper, it is pertinent to note that G. E. Davis has been responsible for the overall organization, tracking, and dissemination of information from the beginning of this project. Thus, any and all errors or oversights concerning the actual classification itself or concerning the rationale behind the choices, the literature reviewed and cited, and the introductory text are the responsibility of J. W. Martin.

## NAMES, DATES, AND THE ICZN

The Introduction section of the fourth edition of the International Code of Zoological Nomenclature (ICZN, 1999:xix) states that the Code "does not fully regulate the names of taxa above the family group." This is, as we understand it, an intentional move designed to allow for some flexibility in establishing higher order taxa. Because of this flexibility, there are different schools of thought for recognizing the names of higher taxonomic categories and for crediting the names and dates of these higher taxa. One school of thought would advocate that a different name (and thus a different person and date) should be used each time the constituency of the taxon is altered. Thus, for example, if the thalassinoid families are removed from the Anomura, then we should no longer use the term Anomura (or use it in a newly restricted sense) to describe the remaining (nonthalassinoid) members of that assemblage. Using another example, if we persist in keeping the taxon name Eumalacostraca and yet exclude the hoplocarids (stomatopods) from the group, we should not credit the name to Grobben, who originally coined the name but considered the hoplocarids to be within the Eumalacostraca. Such changes seem to us to detract considerably from stability and can result in a plethora of new names being proposed for major taxa that essentially have changed very little. An example might be the Achelata of Scholtz and Richter (1995), proposed for what is essentially the Palinura if the family Polychelidae is removed.

The second school of thought maintains that stability is perhaps more valuable than strict accuracy and that there is no need to change (for example) the name Isopoda simply because the tanaidaceans were once included but have since been removed, or to discontinue use of Eumalacostraca because the stomatopods have been removed, or to change the Anomura to Anomala because the thalassinoids have been removed. The latter example was discussed at length by McLaughlin (1983b), who originally advocated using the term Anomala, rather than Anomura, for this reason. Later, McLaughlin and Holthuis (1985) argued for stability and for maintaining the use of the familiar name Anomura. For these reasons, and because the Code reminds us in the Introduction (ICZN, 1999) that "nomenclatural rules are tools that are designed to provide the maximum stability compatible with taxonomic freedom," we side with the second school of thought. Certainly, at lower taxonomic levels, we would never advocate changing the name of a family or genus because of the transfer or synonymy of a single species, and similarly we are hesitant to do away with well-established higher names because their constituency has been slightly altered. Thus, for the most part, we have tended to retain a well-recognized taxonomic name in favor of a new one that differs slightly in its composition.

Another area of controversy is in the crediting of higher taxon names to the original author of the group vs. crediting them to the first person to use the name in its new, higher, context. For example, the ostracode family Darwinulidae is usually credited to Brady and Norman (1889). These authors did not use it to describe any higher taxon, and it was Sohn (1988) who first established the suborder Darwinulocopina (based on this family). Should we refer to the Darwinulocopina Brady and Norman or to the Darwinulocopina Sohn? The ICZN offers some guidelines for resolution of this problem at lower levels via article 50.3.1 (ICZN, 1999:53). This article states that "the authorship of the name of a nominal taxon within the family group, genus group or species group is not affected by the rank at which it is used." This clearly applies only to those mentioned taxonomic levels, and so it does not necessarily need to be invoked for the name of a family that has been elevated to the rank of superfamily (or higher). However, in an attempt to be as consistent as possible, Dr. Lipke Holthuis (who not only is one of the most prolific writers on crustacean systematics in history but also has served on the International Commission of Zoological Nomenclature) has suggested that we extend that recommendation to higher levels for those cases where it was clear to us that the higher taxon had been based on a lower one. Thus, in the above example where the family Darwinulidae has been elevated to superfamily and even to suborder, we might continue to recognize Brady and Norman as the author of both of those higher taxa. Holthuis (1993a) also mentioned ICZN Article 36a (now 36.1), and as an example cited the fact that the "family name Palaemonidae, subfamily name Palaemoninae and the superfamily Palaemonoidea, all have as the author Rafinesque, 1815." The Editorial Preface to the Treatise on Invertebrate Paleontology (Moore, 1969:xi–xxxvi) stated this in a slightly different way, and we quote from it:

> All family-group taxa having names based on the same type genus are attributed to the author who first published the name for any of these assemblages, whether tribe, subfamily, or family (superfamily being almost inevitably a later-conceived taxon). Accordingly, if a family is divided into subfamilies or a subfamily into tribes, the name of no such subfamily or tribe can antedate the family name. Also, every family containing differentiated subfamilies must have a nominate (sensu stricto) subfamily, which is based on the same type genus as that for the family, and the author and date set down for the nominate subfamily invariably are identical with those of the family, without reference to whether the author of the family or some subsequent author introduced subdivisions.

The negative side to following this advice (in the above case, using the taxon names Darwinulidae Brady and Norman and also Darwinulocopina Brady and Norman) is that some "bibliographic" and historical information is lost. The reader will know the original source of the name but will have a very difficult time discovering who first employed that name as a superfamily, suborder, or higher taxon and when this was first done. Using the name "Darwinulocopina Sohn, 1988" is therefore more informative, if not strictly in keeping with ICZN 50.3.1. Holthuis (1993a) was aware of this as well, stating: "One could, in keeping with the rules for the family names, consider the authors of the family name to be at the same time the author of the name of these higher categories, but it seemed more logical to cite as their author the first zoologist who used such a name for a category above the family group level." There are also cases in which the higher taxon was clearly used and described separately, by different authors, rather than being an "elevation" of a family name. For example, within the Peracarida, the family Mictocarididae is correctly credited to Bowman and Iliffe (1985), whereas the order Mictacea is credited to Bowman et al. (1985), who established the order in a companion paper in the same issue of the journal. For these reasons, the choice of author and date following a taxonomic name might at first seem arbitrary, but we have endeavored to credit the person or persons who first used that name in its new (higher) context when this information was known to us. In other instances where we were unsure or where we could not personally check the original literature, we have employed the oldest known name and date, more in keeping with the suggestion by Holthuis (pers. comm.) to extend ICZN 50.3.1 to higher categories. Thus, the present classification, like many others before it, is something of an unfortunate mix of "rules" used to credit authors and dates with the establishment of taxa. M. Grygier (pers. comm.) informs us that the above discussion is slightly misinformed in that the term "family group" explicitly includes superfamilies (ICZN article 35.1), such that the real difficulty should be only at the level of suborder (or any level above that of superfamily).

One of the specific suggestions we received from several workers was a plea to credit Latreille (1803) for a large number of higher level crustacean taxa (we had used the date 1802 in earlier editions of the classification). These taxa include Ostracoda, Malacostraca, Gammaridae (and thus Gammaridea), Oniscidea (and thus Oniscoidea), Astacidea (and thus Astacoidea), Palinura, Paguroidea, Brachyura, Squilloidea, and many more. Our choice of 1802 instead of 1803 is based on the following information quoted from a letter we received from L. Holthuis (pers. comm., 13 July 1998) referring to an earlier draft of our classification:

> Some of Latreille's names proposed in his *Histoire naturelle générale et particulière des Crustacés et des Insectes,* vol. 3 . . . have been cited with the year 1802 . . . others have the year 1803. The year of publication of vol. 3 of Latreille's work was studied by the best authority on Latreille, namely C. Dupuis, who in 1975 (*Bulletin of Zoological Nomenclature,* 32: 4) stated that this vol. 3 was published after April 1802 and before 6 November 1802, thus definitely in 1802. Therefore all the author's names 'Latreille, 1803' should be changed to 'Latreille, 1802.'

Similarly, unless we had fairly convincing evidence to the contrary, in those cases where we were faced with a choice of different dates (which usually, although not always, meant also different authors, such as White, 1850 vs. Dana, 1853 vs. Harger, 1879, all suggested to us by different workers as the correct author/date of the isopod family Limnoriidae) for the establishment of a taxon, we went with the earliest date. In this particular example, at least, it proved the correct choice, as White (1850) is indeed the author of the family Limnoriidae (G. Poore, pers. comm.).

Finally, we wish to caution readers that we have not been able to research each name to the degree that we would have liked, and we have depended instead upon the many contributors (not all of whom were in agreement). Consequently, we would advise any user of this (or any other) classification to take the time necessary to research carefully the history of each taxonomic name for his- or herself, which, because of the sheer number of names involved in this project, we simply were not able to do.

## CLADISTICS AND CLASSIFICATION OF THE CRUSTACEA

Ideally, a classification should accurately reflect the phylogenetic history of the group. We are very much in favor of following rigorous cladistic analyses wherever possible, and some of the newly proposed classification reflects phylogenetic hypotheses based on cladistic analysis of morphological and/or molecular data. However, saying that we favor classifications based on rigorous cladistic methods is not the same as saying that any cladistic analysis is more correct than every preceding hypothesis of crustacean phylogeny. We wish to state this more clearly so that there can be no mistaking our meaning: A phylogeny is not correct simply because it was generated using cladistics. This somewhat obvious point is quite often overlooked. The advantage that cladistics imparts is the objective use of synapomorphies to define clades. Cladistics is a powerful tool, and, like all such tools, it must be wielded carefully. And, as with any other tool, there is never any guarantee that the result is "correct." We received numerous suggestions that we employ a "more cladistic" approach to our new classification. For many crustacean assemblages, there have been no proposed phylogenies, cladistic or otherwise. For other groups, although cladistic methods may have been used, there are no published or accessible data for confirmation of the results, and/or the proposed phylogenies are in stark contrast with large literatures on fossil, morphological, developmental, or molecular studies of these taxa, making them, at least to us, suspect. Two taxa that demonstrate this problem are the Maxillopoda and the Decapoda, for which some of the most vocal proponents of cladistic approaches gave us quite different suggestions for the classification, all supposedly based on rigorous cladistic analyses of "good" data. Similar frustration concerning recent attempts to cladistically analyze fossil arthropods is expressed by Fryer (1999c). More troubling still is that there are other cladistic analyses of which we are aware, and that appear to be based on solid evidence, that we could not follow completely because to do so would have orphaned large numbers of families. For example, we do not doubt the revelation by Cunningham et al. (1992) that king crabs of the family Lithodidae are actually nested within one clade of hermit crabs (but see McLaughlin and Lemaitre, 1997, 2000, for a dissenting opinion). But there are other clades of hermits and other species of lithodids that were not part of this study, and we hesitated to make sweeping changes before all evidence is in. Another example concerns dromiacean crabs, traditionally placed among the lower Brachyura but whose larvae appear distinctly anomuran. The molecular analysis of Spears et al. (1992) grouped at least one dromiid with the Anomura rather than the Brachyura—but does this hold for *all* crabs in the former Dromiacea? Thus, we have in some instances knowingly presented groupings for which contrary evidence exists for at least some of the constituent taxa. We have tried to mention all such areas in the text of the Rationale section that follows. Several workers noted this problem and suggested that perhaps no classification should be attempted until such time that we have better supported phylogenetic analyses in hand for all (or at least most) crustacean groups. There is merit to this argument. But in keeping with our original goal of updating a classification of the entire assemblage to benefit students who wish to view the overall picture of crustacean diversity, we felt that waiting would not improve the situation.

An additional practical problem faced by the student wishing to construct a cladistically based classification is the very real difficulty of representing complex relationships in a two-dimensional classification. To accurately depict all of the branching relationships and show all of the sister groupings would necessitate a rather large number of additional taxonomic categories. One proposed solution is to simply indent the families in the list (without creating additional names for groupings) to imply the relationships. But even this is difficult when dealing with the number of families in, for example, the gammaridean amphipods or the harpacticoid copepods. Another proposed solution is to completely abandon Linnaean hierarchical classifications in favor of a more phylogenetically based system (e.g., see Milius, 1999; Cantino et al., 1999). We feel that, in many cases, a "standard" classification—that is, a simple list of families—still serves a purpose for those taxa where the phylogeny remains uncertain (which is nearly every group of the Crustacea) in that it at least allows recognition and placement within well-defined higher groups for beginning students. Thus, while very much in favor of the application of cladistic methodology and of the construction of classifications based on these methods whenever possible, we have had difficulties in trying to arrive at a sensible or useful way of depicting these relationships to the beginning student of carcinology. Consequently, to many readers, our current arrangements and "lists" of families will appear old fashioned and unsatisfactory.

The number of phylogenetic studies on the Crustacea has risen dramatically since Bowman and Abele's (1982) classification. Christoffersen (1994: 135) estimated that 123 cladistic analyses of crustaceans had appeared in print as of the end of 1992, and that number has increased dramatically since then. Reasons for the increase include improved methods of computation and the availability of cladistic programs, such as PAUP, McCLADE, and HENNIG 86, in addition to the growing acceptance of cladistics as a preferred way of thinking about and depicting crustacean relationships and relationships of all other groups as well (see papers cited in Nielsen, 1995, and Nielsen et al., 1996). Recent phylogenetic software is reviewed by Eernisse (1998), and a list of phylogenetic programs by categories is provided on J. Felsenstein's "Phylogenetic Programs" web site at http://evolution.genetics.washington.edu/phylip/software.html#methods. The fact that cladistics is almost routinely employed in studies of crustacean relationships today can be credited largely to the efforts of F. R. Schram (e.g., see Schram, 1983a, and papers therein; Schram, 1986; Schram and Hof, 1998). Although it is beyond the scope of this project to review the many cladistic analyses of crustacean groups that have appeared since 1982, we list below a few of the more salient papers that treat crustaceans above the level of family, with the hope that this might form something of an introduction to the literature for students of crustacean phylogeny. The list is not intended to be exhaustive. Instead, we hope it alerts readers to the fact that very little is settled with regard to crustacean relationships and classification and to the fact that cladistic thinking has profoundly affected our understanding of crustacean relationships.

In alphabetical order within chronological order, these works include: Briggs (1983, Cambrian arthropods and crustaceans [see also Briggs and Whittington, 1981]), Grygier (1983a, b, maxillopodans), Sieg (1983a, tanaidaceans), Takeuchi (1993, caprellidean amphipods), Wheeler et al. (1993, arthropods including crustaceans), Ho (1984, nereicoliform copepods), Schram (1984a, Eumalacostraca; 1984b, Syncarida), Martin and Abele (1986, anomuran decapods), Schram (1986, all crustacean groups), Christoffersen (1986, 1987, caridean shrimp), Grygier (1987a, b, maxillopodans), Pires (1987, peracarids), Christoffersen (1988a, b, caridean shrimp), Müller and Walossek (1988, Maxillopoda), Abele et al. (1989, pentastomids), Boxshall and Huys (1989a, maxillopodans), Briggs and Fortey (1989, Cambrian arthropods including crustaceans), Christoffersen (1989, caridean shrimp), Schmalfuss (1989, oniscidean isopods), Brusca and Brusca (1990, all crustacean groups), Christoffersen (1990, Caridea), Ho (1990, copepod orders), Kim and Abele (1990, decapods), Walossek and Müller (1990, "stem line" crustaceans), Abele (1991, decapods), Brusca and Wilson (1991, isopods), Abele et al. (1992, maxillopodan groups), Briggs et al. and Briggs and Fortey (1992, Cambrian arthropods including crustaceans), Høeg (1992a, maxillopodans), Spears et al. (1992, brachyuran crabs), Walossek and Müller (1992, "orsten" fossil crustaceans), Wilson (1992, most major extant groups), Kim and Kim (1993, gammaridean amphipod families and amphipod suborders), Walossek (1993, branchiopods and Crustacea), Poore (1994, thalassinideans), Spears et al. (1994, thecostracan maxillopodans), Wagner (1994, peracarids), Wilson (1994, janiroidean isopods), Glenner et al. (1995, cirripedes), Scholtz and Richter (1995, decapods), Bellwood (1996, calappid crabs), Humes and Boxshall (1996, lichomolgoid copepods), Moura and Christoffersen (1996, "mandibulate" arthropods), Wilson (1996, isopods), Ahyong (1997, stomatopods), Emerson and Schram (1997, all arthropods), Hanner and Fugate (1997, branchiopods), Spears and Abele (1997, several major groups, review), Tshudy and Babcock (1997, clawed lobsters), Tudge (1997b, anomurans), Walossek and Müller (1997, Cambrian crustaceans and their bearing on crustacean phylogeny), Wheeler (1997, arthropods including crustaceans), Wills (1997, all Crustacea), Jenner et al. (1998, hoplocarids), Olesen (1998, conchostracans and cladocerans), Schram and Hof (1998, all major groups, extant and extinct), Shen et al. (1998, spelaeogriphaceans), Strausfeld (1998, crustacean neurological features), Taylor et al. (1998, mysidaceans and other peracarids), Tucker (1998, raninoid crabs), Wheeler (1998, all arthropod groups), Wills et al. (1998, fossil and extant arthropod groups), Almeida and Christoffersen (1999, pentastomids), Cumberlidge and Sternberg (1999, freshwater crabs), Huys and Lee (1999, laophontoidean harpacticoid copepods), Sternberg et al. (1999, freshwater crabs), Olesen (1999b, leptostracans), Spears and Abele (1999b, crustaceans with foliaceous limbs; 2000, branchiopods), Walossek (1999, major crustacean groups), Edgecomb et al. (2000, all major arthropod groups), Negrea et al. (1999, branchiopods), Shultz and Regier (2000, all major arthropod groups), and Richter et al. (2001, cladocerans).

## MOLECULAR SYSTEMATICS AND CLASSIFICATION OF THE CRUSTACEA

Without doubt, the most exciting recent developments in our understanding of crustacean relationships have been in the realm of molecular systematics and phylogenetics. Indeed, many of the cladistic papers mentioned in the previous section are based on molecular sequence data, which essentially were not available at the time of the Bowman and Abele classification. Molecular systematic studies of arthropods have become so numerous that Wheeler (1998) stated "the past decade has presented us with nearly annual molecular analyses of Arthropoda." For the Crustacea, most of this work has been championed by the laboratories of L. G. Abele and T. Spears at Florida State University and C. W. Cunningham at Duke University. This field, as well as the field of developmental genetics (which we barely touch upon here), is growing and changing at a phenomenal rate. Many of the early studies were based on relatively small sequences, so it is not terribly surprising that there have been some published results that appear unreasonable based on our knowledge of morphology, embryology, paleontology, and other sets of characters. As we refine our selection of which genes to target, improve our ability to extract and align increasingly larger sequences, and devise better computational algorithms, we might begin to see more agreement between molecular results and more traditional views of crustacean phylogeny, or at least results that are less ambiguous. Or we may not. As Spears and Abele (1997) state in the conclusion to their review paper on the use of 18S rDNA data in crustacean phylogeny, "Regrettably, in the crusade for understanding relationships among crustaceans and other arthropod lineages, the rDNA data represent but a relic, and not the Holy Grail itself." Yet despite this sobering conclusion, Spears and Abele (1997) were able to make some very strong statements concerning at least some crustacean taxa. For example, Branchiopoda, Copepoda, Podocopida, and Myodocopida are all clearly monophyletic; the Malacostraca is clearly monophyletic and includes the Phyllocarida (Leptostraca) (supported also by Shultz and Regier, 2000); Maxillopoda does not appear monophyletic (although certain groups within it seem to be united); etc.

There are, of course, known problems associated with some of these approaches (as one early example, see the responses by Nielsen and others (1989) to the article by Field et al. (1988) entitled "Molecular analysis of the animal kingdom"). Fryer (1997) points out several papers that question the results and/or validity of recent studies of arthropod phylogeny based on molecular data; Wägele and Stanjek (1995) make the point that alignment alone can be responsible for serious discrepancies in analyses of such data. And of course the history of a particular gene might not accurately reflect the phylogeny of the species containing that gene (e.g., see Brower et al., 1996; Doyle, 1997; Maddison, 1997; Page and Charleston, 1998). Unfortunately, the branchiopod genus *Artemia,* which has been used for more molecular comparative studies than any other crustacean genus, is not the best choice; Maley and Marshall (1998) note that "brine shrimp [have] long been known to produce artifactual groupings." Lake (1990) admitted that arthropod paraphyly as indicated in his analysis may be a result of long branch attraction caused by the inclusion of *Artemia* and *Drosophila*; this problem was mentioned also by Turbeville et al. (1991). It is also disconcerting that, after so much money and effort have been expended toward applying genetic data to resolving the evolutionary roots of modern humans, we still do not have a clear answer. Whether *Homo sapiens* arose from a single African source 200,000 years ago or "multiple groups in Africa and elsewhere" at least a million years ago is still hotly debated (see Bower, 1999). How, then, are we expected to place confidence in what the molecules are telling us about the evolution of crustaceans when our efforts, in comparison, have been so limited? To summarize, we again quote Maley and Marshall (1998): "To be confident in our hypotheses of relationships among the animal phyla we need to gather more DNA sequences, especially from undersampled phyla; develop better methods of DNA analysis on the basis of more realistic models of DNA evolution; and develop independent data sets using morphological, developmental, and other molecular data to corroborate or falsify specific hypotheses or to combine in total-evidence analyses." Thus, just as we have not accepted all cladistic analyses simply because they were cladistic, we have incorporated molecular analyses with caution because of perceived problems with some of these studies. At the same time, there is little question that these efforts, however preliminary they may be, represent the first attempts to apply "new" and objective data to the resolution of crustacean phylogeny for the first time in some 200 years of study, and we look forward to continued advances in this field.

Papers mentioned below are merely examples of some of the more comprehensive or influential works of which we are aware. As in the previous section, we have included only those papers that deal with "higher level" crustacean taxa or with the relationships of crustaceans to other arthropods. In alphabetical order within chronological order, these papers include Abele et al. (1989, pentastomids, rRNA), Kim and Abele (1990, decapods, 18S rRNA), Abele (1991, decapods, 18s rRNA), Turbeville et al. (1991, arthropods including crustaceans, 18S rRNA), Abele et al. (1992, maxillopodans, 18S rDNA), Cunningham et al. (1992, lithodid and pagurid anomurans), Spears et al. (1992, brachyuran crabs, 18s rRNA), Wheeler et al. (1993, arthropods including crustaceans, 18S rDNA, and polyubiquitin), Raff et al. (1994, review of arthropod relationships [and other metazoan groups] based on various genes), Spears et al. (1994, thecostracans, 18S rDNA), Boore et al. (1995, arthropods including crustaceans), Friedrich and Tautz (1995, arthropods, 18S and 28S rDNA), France and Kocher (1996, DNA sequencing of formalin-fixed crustaceans), Wray et al. (1996, 6 mitochondrial and 2 nuclear genes), Eernisse (1997, arthropods [including crustaceans] and annelids, 18S rRNA), Hanner and Fugate (1997, branchiopods, 12S rDNA), Regier and Schultz (1997, major arthropod groups, two nuclear genes), Spears and Abele (1997, all crustacean groups, 18S rDNA), Wheeler (1997, most arthropod groups), Boore et al. (1998, crustaceans and insects, gene translocations), Colgan et al. (1998, arthropods including crustaceans, histone H3 and U2 snRNA), Min et al. (1998, arthropods, 18S rDNA), Regier and Schultz (1998a, b, arthropods, amino acid sequence of EF-1α), Schwenk et al. (1998, cladocerans, 16S rDNA), Wheeler (1998, arthropods [including crustaceans], 18S and 28S rDNA), Braga et al. (1999, copepods, 16S and 28S rRNA), Morrison and Cunningham (1999, anomurans, mitochondrial gene rearrangements), Spears and Abele (1999b, crustaceans with foliaceous limbs, 18S rDNA), Crandall et al. (2000, Astacidea, 18S, 28S, and 16S rDNA), Edgecomb et al. (2000, arthropods including crustaceans, histone H3 and U2 snRNA sequences), Giribet and Ribera (2000, all arthropod groups, 18S and 28S rDNA), Harris et al. (2000, barnacles, 18S rDNA), Jarman et al. (2000, malacostracans, 28S rDNA), Perl-Treves et al. (2000, thecostracans, 18S rDNA), Remigio and Hebert (2000, anostracan branchiopods, 28S and 16S rDNA), Spears and Abele (2000, branchiopods, 18S rDNA), Schubart et al. (2000a, b, grapsoid crabs, 16S rDNA), Shultz and Regier (2000, arthropods, Ef-1α and Pol II), Wilson et al. (2000, Malacostraca, mitochondrial DNA and gene order), Mattern and Schlegel (2001, oniscidean isopods, ssu rDNA), and Richter et al. (2001, Cladocera, 12S rDNA). See also papers in the symposium *Evolutionary Relationships of Metazoan Phyla* organized by D. McHugh and K. Halanych (1998, *American Zoologist* 38:813–982) and the volume *Arthropod Relationships* edited by R. A. Fortey and R. H. Thomas (1997).

## DEVELOPMENTAL GENETICS AND CLASSIFICATION OF THE CRUSTACEA

The relatively newly emerging field of developmental genetics needs to be mentioned here as well, though we hasten to add that this field of study is well beyond our area of expertise and that any attempt at a synthesis would be premature. Recent discoveries concerning especially homeotic (Hox) genes and arthropod relationships are having a profound influence on our understanding of crustacean morphological plasticity and clearly will play an increasingly important role in elucidating relationships within Crustacea and among the various arthropod groups. We include this brief section only as a way to signal to the beginning student what is surely to be an active field of research for many years to come. Some of the recent papers in this field with applications to crustacean classification include (in alphabetical order) Akam (1998), Akam et al. (1994), Arhat and Kaufman (1999), Averof and Akam (1993, 1995a, b), Averof and Patel (1997), Carroll (1995), Davidson et al. (1995), Fortey and Thomas (1997), Grenier et al. (1997), Panganiban et al. (1995, 1997), Popadić et al. (1996), Roush (1995), Scholtz (1995), Shubin et al. (1997), and Williams and Nagy (1995) (some of which are briefly reviewed in Brusca, 2000).

## SPERM MORPHOLOGY AND CLASSIFICATION OF THE CRUSTACEA

Yet another field of research that is improving our understanding of crustacean relationships is the description and comparison of crustacean sperm, termed "spermiocladistics" by Jamieson (1987, 1991a). While examination of crustacean sperm morphology for systematic purposes is not new (e.g., Koltzoff, 1906; Wingstrand, 1972, 1978, 1988; Grygier, 1981, 1982), recent work has employed ultrastructural characters that show more promise for resolution of long-standing questions. In the words of Tudge (1997b), the "use of spermatozoal ultrastructure in taxonomy and phylogeny is now firmly established as a valid means of investigating phylogenetic relationships in various animal phyla." For the Crustacea, these characters have been invoked mostly for resolving relationships within the Eumalacostraca. This work is being championed primarily by B. G. Jamieson and C. Tudge and their colleagues. Some of the many recent papers advocating sperm ultrastructural characters in phylogeny are Guinot et al. (1994, primitive crabs; 1997, freshwater crabs; 1998, dromiacean crabs), Richer de Forges et al. (1997, crabs), Jamieson (1989a, b, crabs; 1989c, stomatopods; 1990, primitive crabs; 1991a, overview of crustacean sperm ultrastructure and phylogeny; 1991b, 1993, 1994, crabs), Jamieson et al. (1993a–c, crabs; 1994a, b, 1995, 1996, 1997, crabs), Jamieson and Tudge (1990, crabs), Jamieson, Tudge, and Scheltinga (1993, primitive crabs), Jespersen (1979, leptostracans), Grygier (1981, 1982, maxillopodans), Storch and Jamieson (1992, pentastomids), Tudge (1991, 1992, 1995, 1997a, b, 1999a, b, anomuran decapods), Tudge et al. (1998a, lithodid crabs; 1998b, hydrothermal vent crabs), and Tudge et al. (2000, mud-shrimp families; 1999, hippoid crabs). Many of these papers and their contributions are discussed in the sections dealing with the taxa in question.

Some of the revelations from the study of sperm ultrastructure are not terribly surprising and in fact support previous long-standing hypotheses of crustacean relationships (e.g., peracarid unity; Jamieson, 1991a). Other results are more controversial and include the alliance of the Remipedia with the Maxillopoda on the basis of the shared "flagellate condition" of their spermatozoon (Jamieson, 1991a) and placing the genus *Lomis* outside of, and thalassinids within, the Anomura (Tudge, 1997a, b) (in contrast with what Morrison and Cunningham, 1999, presented based on mitochondrial gene rearrangement data). [As an aside, the congruence between the phylogenetic diagrams of Jamieson (1991a:111), based on sperm ultrastructure, and Schram (1986), based on cladistic analysis of morphological characters, is perhaps not so remarkable as Schram and Hof (1998) suggest. Schram and Hof (1998) refer to Jamieson's figure and ask the reader to "note the general correspondence with the major classes as arranged in Fig. 6.1.A." However, Jamieson's figure was in turn based on Schram (1986) with a diagram of the spermatozoal ultrastructure simply added to Schram's tree; it is not an independently derived phylogeny.] Continued use of sperm ultrastructure in crustacean taxonomy and systematics will almost certainly contribute significantly to our understanding of crustacean phylogeny.

## LARVAL MORPHOLOGY AND CLASSIFICATION OF THE CRUSTACEA

The study of crustacean systematics and phylogeny has involved larval characters from the very earliest times. For many groups of crustaceans, a study of systematic relationships *is* a study of the larvae, as these are often the only characters, or the best characters, that we have. For example, it could be argued that, until recently, the history of studies in barnacle phylogeny has been essentially a history of comparisons of barnacle larvae, and to some extent this is true for many groups. For some taxa, in particular the Facetotecta, the larvae are all that we know; the adult has yet to be recognized or described. The reverse is also true: there are still some important groups of crustaceans (the class Remipedia, for example) for which the larval forms have never been identified. Many of the classic treatments of crustacean larvae were published prior to the Bowman and Abele (1982) classification and were thus available for consideration by those authors. The summary of crustacean larval diversity published by Williamson in that same series of volumes (Williamson, 1982) remains a good entry point for the literature on crustacean larvae and relationships based on larval characters.

In the years following the Bowman and Abele (1982) classification, there have been additional and significant treatments of crustacean larval characters and phylogeny. Indeed, nearly every modern publication that describes a larval stage includes at least some comments on the applicability of the findings to relationships within the group. The study of larval crabs, in particular, has been a rich source of new characters for postulating higher level relationships among the Brachyura (e.g., see Rice, 1980, 1981, 1983, 1988; Martin, 1984, 1988; Martin et al., 1985; Felder et al., 1985, as a few selected examples from a huge body of literature on crab relationships based on larvae and postlarvae). Williamson (1988a, b) has proposed rather drastic changes in our understanding of various pleocyemate groups (particularly the position of the dromiid crabs relative to anomurans and true crabs, the placement of the mysidaceans within the Eucarida, and the separation of palinurid lobsters from other eucarids based on their bizarre larvae). Grygier (1987a–c) and others have used larval characters to explore maxillopod phylogeny; within the Maxillopoda, the work of Dahms (e.g., Dahms, 1990) could be mentioned for advancing our understanding of copepod naupliar characters in phylogeny. Discoveries of fossilized larvae, in particular papers on the "Orsten" fauna, have added new characters and new insights into the evolution of early crustaceans and "stem-line" crustaceans (e.g., see Müller and Walossek, 1985a, 1986b; Walossek, 1993, 1995; Walossek and Müller, 1990, 1997). Walossek and Müller (1997) recognize the Entomostraca, and exclude from the Crustacea the Pentastomida, in part based on larval evidence.

We have tried to mention studies based on larval characters (where they have a bearing on classification at the family level or higher) under each crustacean taxon. A recent review of larval diversity (Harvey et al., in press) provides additional material geared primarily for the beginning student of carcinology.

## THE FOSSIL RECORD AND CLASSIFICATION OF THE CRUSTACEA

No understanding of crustacean diversity and evolution would be complete without knowledge of the fascinating fossil history of the group. And many exciting discoveries that bear on crustacean origins, relationships, and classification have surfaced since the Bowman and Abele treatment. A recent example is the intriguing find of a serolid-like sphaeromatoid isopod from the Solnhofen of Germany (Brandt et al., 1999), pushing back the origin of sphaeromatoid isopods to at least the Late Jurassic. Although a thorough review of such discoveries is beyond the scope of this report (see papers in Edgecombe, 1998, and reviews by Delle Cave and Simonetta, 1991; Bergström, 1992; Schram and Hof, 1998; Walossek and Müller, 1997, 1998; Wills, 1998; Wills et al., 1995; Fortey et al., 1997; Fryer, 1999c), we feel the need to mention especially the stem and crown group crustaceans of the "Orsten" fauna of Sweden (Orsten-type fossils have also been found on other continents; see review by Walossek, 1999). These works include papers by Müller (1982, *Hesslandona*; 1983, crustaceans with soft parts), Müller and Walossek (1985, Skaracarida; 1986a, *Martinssonia*; 1986b, various arthropod larvae; 1988, the maxillopod *Bredocaris*), Walossek and Müller (1990, stem line crustacean concept; 1992, overview of the Orsten fauna; 1994, possible pentastomids; 1997, 1998, overviews), Walossek and Szaniawski (1991, *Cambrocaris*), Walossek et al. (1994, possible pentastomids), and Walossek (1993, 1995, the branchiopod *Rehbachiella*; 1999, overview of Cambrian crustaceans). These publications include detailed descriptions of several new taxa that have in many ways altered our view of primitive crustaceans and the timing of crustacean evolution.

The Burgess Shale crustaceans have been reexamined recently by Briggs et al. (1994), and the remarkable fossil arthropods from the Lower Cambrian Chengjiang fauna of southwest China have been summarized by Hou and Bergström (1991, 1997). Included in the Chengjiang fauna are no unequivocal crustaceans (*Waptia* being the only remote possibility), but several fossils seem to have a bearing on our understanding of crustacean evolution. Other recent studies of Chinese fossil crustaceans have included papers on conchostracans (e.g., Shen, 1984, 1990; Zhang et al., 1990; see also Orr and Briggs, 1999, for Carboniferous conchostracans from Ireland), and Lower Cambrian crustaceans are known from other sites around the world as well (e.g., see Butterfield, 1994). Studies of bradoriid and phosphatocopid arthropods (once thought to be ostracodes) (see Siveter and Williams, 1997) have even shed light on our understanding of the evolution of the crustacean circulatory system (Vannier et al., 1997). The phosphatocopids are now thought to be close to the "stem-line" crustaceans (and possibly the sister taxon to Crustacea; see Walossek, 1999) rather than relatives of any of the crown-group crustaceans such as ostracodes or maxillopods, which had been suggested previously (e.g., see reviews by Walossek and Müller, 1992, 1998). At least two major groups, and possibly many more unknown to us, remain enigmatic as to whether they belong in the Crustacea or not: Thylacocephala (see Pinna et al., 1982, 1985; Secretan, 1985 [as Conchyliocarida]; Rolfe, 1985, 1992; Schram et al., 1999) and Cycloidea (see Schram et al., 1997; Schram and Hof, 1998), although cycloids were probably allied to the maxillopodans (Schram et al., 1997). Schram and Hof presented, as part of the Fourth International Crustacean Congress (ICC-4) in Amsterdam, evidence that the Thylacocephala are indeed crustaceans; they further postulate the inclusion of the Thylacocephala in the Thecostraca on the basis of the presence of lattice organs. Their paper, entitled "At last: the Thylacocephala are Crustacea," was a late addition and therefore is not included among the published abstracts of the ICC-4 Congress, but since then, the information has been submitted (Lange et al., in press). However, Schram et al. (1999) are more cautious and stopped short of declaring that thylacocephalans were crustaceans. The Permian "pygocephalomorph" crustaceans and their relationship to extant mysidaceans was examined recently by Taylor et al. (1998). A thorough review of most of the above contributions is presented by Schram and Hof (1998). Many other papers on crustacean fossils continue to add to our knowledge of the history of the group (e.g., Brandt et al., 1999, on the Late Jurassic origin of sphaeromatoid isopods).

In light of these remarkable finds, it is understandable that a number of colleagues have suggested, some rather strongly, that we incorporate fossil taxa into the current classification. We have opted not to do so, primarily because we are less familiar with the fossil crustacean literature (and with workers in that field) than we are with the literature on extant groups. Thus, the opportunities for us to inadvertently perpetuate or create errors would have been much greater had we attempted this task. Also, if the currently proposed classification proves to have merit, it should not be difficult for more paleontologically inclined carcinologists to, at some point, add these fossil taxa to the existing framework. We hope that the classification is of some use to paleontologists and that, at some point, we can incorporate fossil taxa into this scheme. Relatively recent lists of crustacean fossil taxa can be found in Whatley et al. (1993, ostracodes) and Briggs et al. (1993, all other crustacean groups) (both in M. J. Benton, editor, *The Fossil Record 2,* Chapman and Hall, 1993). However, since our knowledge (and time) is limited, we have decided to include only extant taxa for now.

## A NOTE ON THE APPENDICES

## APPENDIX I. COMMENTS AND OPINIONS

After receiving and considering the input from various workers around the world, we then asked the same persons to comment on the resulting product. We did this for two reasons. First, many of the suggestions we received were not incorporated, and we wanted collaborators to have the opportunity to express their disagreement. Reasons for not incorporating a particular suggestion were many and ranged from simple disagreement on our part to conflicting suggestions or corrections from noted experts. Second, we wanted students of carcinology to know where the major areas of disagreement lie in our understanding of crustacean phylogeny and classification. By pointing out areas where other experts in the field disagree with the current classification, we hoped to avoid the impression that the classification is accepted or agreed upon by some consensus of crustacean taxonomists.

## APPENDIX II: LIST OF CONTRIBUTORS

The list of persons to whom we sent either first, second, or third drafts of the classification is given in Appendix II. Some of those listed responded to only one of our mailings; some responded to all mailings; some workers did not respond at all. No person on the list should be assumed to be in agreement with the classification as a whole. Despite these caveats, we felt that we should list all of the workers we attempted to contact to let readers know the potential pool of expertise from which we solicited input.

Because of the tremendous interest in the Crustacea worldwide, the number of qualified workers is much greater than this list indicates. Our decision on whose input to solicit was more or less arbitrary, based on our own knowledge of workers in the field and on suggestions received as a result of the first and second mailings. We apologize in advance if, by omitting someone from one or more mailings, we have inadvertently slighted anyone; such was not our intent.

## APPENDIX III: OTHER CRUSTACEAN RESOURCES

Finally, a list of other crustacean resources is provided to give the student of Crustacea an introduction to the large and ever growing number of crustacean resources. The list includes crustacean-specific journals, newsletters of special interest groups (e.g., *Zoea, Ecdysiast, Monoculus, Anostracan News,* and *Cumacean Newsletter*), and URLs of helpful crustacean-related sites on the World Wide Web.

# RATIONALE

## SUBPHYLUM CRUSTACEA

Many of the questions considered most pressing today have been asked for well over 100 years: Are crustaceans a monophyletic group? How many major clades, or classes, are there? Which is the most primitive class? What are the relationships among the classes? We cannot attempt to answer all of these questions here, but below we offer a brief explanation of how and why we arrived at the current classification. In most cases, we provide some additional information under the heading for each of the various taxa (each of which is treated later). For more in-depth discussions of the complex history of attempts to classify the Crustacea, we refer the reader to the following publications: Moore and McCormick (1969), Schram (1986), Spears and Abele (1997), Schram and Hof (1998), and especially Monod and Forest (1996).

### Are Crustaceans a Monophyletic Group?

The question of crustacean monophyly, the place of the Crustacea within the Arthropoda, the question of arthropod monophyly, and the relationships among the many arthropod and crustacean groups have been reviewed by several recent workers (see especially Boore et al., 1995; Friedrich and Tautz, 1995; Telford and Thomas, 1995; Raff et al., 1994; Fortey et al., 1997; Regier and Shultz, 1997, 1998b; Wheeler, 1998; Shultz and Regier, 2000; Edgecomb et al., 2000). Broader questions concerning whether crustaceans and other arthropods belong in a phylum or larger clade called the Ecdysozoa (see Garey et al., 1996; Aguinaldo et al., 1997) are reviewed by Schmidt-Rhaesa et al. (1998) and Garey (2000). We have not attempted to address either of these issues (that is, the relationship of crustaceans to other arthropods or the relationships within the Ecdysozoa) and instead refer the reader to the following publications and the papers cited therein. Wheeler et al. (1993) presented a combined analysis of morphological and molecular data that strongly supported arthropod monophyly, and this view was strengthened by Wheeler (1998). Lake (1990) suggested arthropod paraphyly, while Fryer (1997) presents several arguments in favor of arthropod polyphyly. Strausfeld (1998) depicts insects and crustaceans (both of which he feels may be paraphyletic) as sister groups on the basis of neuroanatomical data. Preliminary work on the neurogenesis of compound eyes supports common ancestry for crustaceans and insects as well (e.g., see Harzsch and Walossek, 2001, and references cited therein). Friedrich and Tautz (1995) support both arthropod monophyly and a crustacean–insect sister group arrangement with DNA sequence data, as do Boore et al. (1995, 1998), using mitochondrial gene rearrangement data, and Wilson et al. (2000), comparing the complete mitochondrial genome of a malacostracan with that of *Drosophila.* Regier and Schultz (1997, 1998a, b) also questioned crustacean monophyly (their 1997 title suggests crustacean polyphyly) based on EF-1$\alpha$ and RNA polymerase II (Pol II); however, their results were somewhat ambiguous, as there were no strongly supported nodes, and support for a basal Malacostraca was not high (J. Regier, pers. comm.). Regier and Shultz also suggested (1997, 1998b), as had other workers, that branchiopod crustaceans may be more closely related to other arthropod groups (hexapods and myriapods) than they are to malacostracan crustaceans, although this too did not have strong node support (what was strongly supported was that branchiopods, and indeed all of our six classes of crustaceans, grouped with hexapods to the exclusion of myriapods, arguing against the concept of the "Atelocerata" (hexapods + myriapods); see also Popadić et al., 1996, and Shultz and Regier, 2000). Another way of stating this is that, if crustaceans are not monophyletic, then the group that breaks them up is the Hexapoda and not myriapods or chelicerates or groups outside Arthropoda. The emerging field of developmental biology (see references cited in the earlier section on developmental genetics and crustacean classification) also provides evidence that crustaceans and insects are closely linked. Brusca (2000) nicely summarizes the history of the controversy and the disparate data sets. Two recent volumes address these questions by way of collections of edited papers: Fortey and Thomas (1997, *Arthropod Relationships,* Chapman and Hall) and Edgecombe (1998, *Arthropod Fossils and Phylogeny,* Columbia University Press).

In the introduction to the latter volume, Edgecombe notes that "the monophyly of Crustacea is endorsed in every chapter that investigates the issue" (see also Edgecomb et al., 2000). Yet there remains some doubt. We have found it advantageous, at least for the project at hand, to treat the group as monophyletic. We also note that there is an abundance of fossil, morphological, and molecular data that support this view. The "crown-" vs. "stem-group" approach as detailed by Walossek and Müller (1990, 1998) is worth noting in this regard; those authors consider the Crustacea monophyletic and give several morphological characters that uniquely define the group, while at the same time they present interesting information on "stem-line crustaceans," crustacean-like arthropods that are not members of the crown group (their "Eucrustacea") but that share at least some features with true crustaceans. Other workers have argued, some with more data than others, that the Crustacea is paraphyletic (e.g., Moura and Christoffersen, 1996; García-Machado et al., 1999; Wilson et al., 2000) or polyphyletic (e.g., Averof and Akam, 1995a, b) or that the question is, at best, unresolved (e.g., Regier and Schultz, 1997, 1998a, b; Shultz and Regier, 2000), and we would be remiss not to mention these dissenting opinions. Further arguments for or against the monophyly of the Crustacea (and also Arthropoda) can be found in the reviews by Brusca (2000) and Giribet and Ribera (2000).

Our treatment of the Crustacea as a subphylum (of the Arthropoda) is therefore somewhat arbitrary. Arguments could be (and have been) made for recognizing the group as a distinct phylum, and some workers refer to the Crustacea as a superclass or class. Our choice of subphylum allowed us to use classes within the group, which to us was more manageable. Treating the Crustacea as a subphylum implies monophyly of the Arthropoda. Although this issue is not completely settled (see above references and especially Fryer, 1997, in Fortey and Thomas, 1997), most bodies of evidence of which we are aware seem to indicate that the arthropods are indeed a phylum (see summaries in Raff et al., 1994; Telford and Thomas, 1995; and Brusca, 2000) that includes the Crustacea.

## How Many Classes Are There?

The history of higher level classification of the Crustacea is briefly discussed in Holthuis (1993a), Spears and Abele (1997), Schram (1986), Schram and Hof (1998), and especially Monod and Forest (1996). Some of the more notable schemes for crustacean classification that have appeared subsequent to the Bowman and Abele (1982) classification are those of Schram (1986), Starobogatov (1986, with English translation by Grygier in 1988), and Brusca and Brusca (1990). Other workers have presented phylogenies from which the reader can deduce alternative classifications, even if no specific classification is presented in the paper (e.g., Wilson, 1992).

Schram (1986) departed from Bowman and Abele's use of six classes by recognizing four groups: Remipedia, Phyllopoda (which included the branchiopods, cephalocarids [as Brachypoda], and leptostracans), Maxillopoda (including tantulocarids, branchiurans, mystacocaridans, ostracodes, copepods, facetotectans, rhizocephalans, ascothoracidans, acrothoracicans, and thoracicans), and Malacostraca (containing both the hoplocarids and the eumalacostracans). Schram's (1986:542–544) classification extends to the level of suborder and occasionally infraorder. It is noteworthy not only for attempting to derive a classification from his cladistic analyses but also because of his inclusion of a large number of fossil taxa. Unfortunately, Schram (1986) also introduced, or employed, some taxonomic names that have not been well accepted (e.g., "Euzygida" for the stenopodidean shrimps; "Eukyphida" for the carideans; "Edriophthalma" to contain the isopods and amphipods as distinct from all other peracarids, etc.). Starobogatov (1986, 1988) recognized four groups as well, but the composition of his four groups differs appreciably from those of Schram and from those of all other previous workers. Additionally, Starobogatov employed some unusual names for his groupings (such as Carcinioides for the malacostracans and Halicynioides to accommodate some of the maxillopodan groups) that are unlikely to receive wide recognition, and his classification appears to be at odds with most of the morphological and fossil data (e.g., see Schram and Hof, 1998) as well as with the molecular data (e.g., Spears and Abele, 1997). Brusca and Brusca (1990) recognized five classes (Remipedia, Branchiopoda, Cephalocarida, Maxillopoda, and Malacostraca), and in part because this usage is in a major textbook, it has received wide acceptance. Bousfield and Conlan (1990, *Encyclopaedia Britannica*), whose classification extends only to the ordinal level, followed Schram's lead for some groups of the Crustacea and Bowman and Abele (1982) for others. Their classification is noteworthy because of their attempt to include fossil taxa as well and because of their laudable attempt to estimate the number of families in each order. Gruner (1993) treats the Crustacea as a class, does not recognize the Branchiopoda or Maxillopoda, and as a result includes 13 separate subclasses. Apart from the somewhat unusual treatment by Starobogatov, the number of proposed or recognized classes seems to have depended mostly upon whether the maxillopods are seen as a natural assemblage and, if they are, whether the ostracodes are within or outside of the Maxillopoda, and on whether and how the Malacostraca should be divided.

In our classification, the subphylum Crustacea includes six major groups, which we are treating as classes: Branchiopoda, Remipedia, Cephalocarida, Maxillopoda, Ostracoda, and Malacostraca. However, this is somewhat misleading in that we are also positing the Branchiopoda as the sister taxon to all other crustacean groups. Thus, the "class" Branchiopoda should be accorded more weight than the remaining classes, which together constitute the sister group to the branchiopods in our arrangement. Our treatment of crustaceans as being comprised of six classes is quite conservative and follows essentially the Bowman and Abele (1982) classification. Perhaps the most salient problem is our continued recognition of the Maxillopoda as a valid class, when virtually all lines of evidence point to its being an artificial assemblage (see discussion under Maxillopoda). Thus, Wilson (1992) observed that "the concept of the Maxillopoda is not supported in any of the trees" and Spears and Abele's (1997) molecular analysis "fails to provide strong support for a monophyletic Maxillopoda." If we eliminated the Maxillopoda as a class, as has Gruner (1993) (and there are many lines of evidence that suggest that this is the correct course), then we would treat as distinct classes each of the currently recognized "maxillopodan" subclasses (the Thecostraca, Tantulocarida, Mystacocarida, and Copepoda). This would have the advantage of further increasing our perception of crustacean diversity (only because nine classes sounds more diverse than six). The number of crustacean classes that should be recognized is a very controversial topic, and opinion is sharply divided. As Spears and Abele (1997) noted, "surprisingly, there is as yet no consensus regarding even the number of constituent crustacean classes."

We do not recognize the taxon "Entomostraca," which has been used historically by several workers in slightly different contexts (e.g., McKenzie et al., 1983; Walossek and Müller, 1998). Walossek and Müller (1998:210) and Walossek (1999) recognize this group as one of the "two major lineages" of Crustacea (the other being the Malacostraca). Contained in their Entomostraca are the cephalocarids (depicted as the sister taxon to the Maxillopoda and Branchiopoda) and two extinct groups (Orstenocarida and Skaracarida).

**Which Is the Most Primitive Class?**

We are treating the class Branchiopoda as the most primitive of the extant groups of Crustacea. We arrived at this decision mostly because of the following three lines of evidence. First, the group as a whole is ancient and extends back into the Upper Cambrian and probably further (see Fryer, 1999, and especially Walossek, 1993). A beautifully preserved fossil from the Upper Cambrian of Sweden (*Rehbachiella*) appears to be a branchiopod and is similar in many ways to living anostracans (Walossek, 1993; although note that Olesen (1999a) questions the anostracan affinities of *Rehbachiella,* while both Wills (1997) and Schram and Hof (1998) obtained nonbranchiopod positions for *Rehbachiella* on their cladograms). There are no known fossils of any cephalocarids, and the only fossils thought to be remipedian are from the Carboniferous (Mississippian and Pennsylvanian) Period (Schram and Hof, 1998). In fairness, we should state also that (1) cephalocarids, because of their habitat, size, and fragility, would seem unlikely candidates for fossilization (and yet, such could also be said about the minute animals in the Orsten fauna) and (2) there are other crustacean groups known from the Upper Cambrian, such that appearance of branchiopods in the Upper Cambrian is not in itself sufficient to argue for their being the most primitive of the extant classes. Second, there are developmental studies that show clear and unambiguous anamorphic development in at least some branchiopods, which is exhibited by no other living crustacean group (e.g., see Fryer, 1983). On the other hand, cephalocarids exhibit only slightly metamorphic development, and as of this writing, we still know nothing about remipede development. Third, some studies based on molecular sequence data seem to indicate that branchiopods are not only monophyletic but are also distinct from all other crustacean assemblages (e.g., Spears and Abele, 1997, 2000; Regier and Schultz, 1997, 1998a, b; Shultz and Regier, 2000). As noted earlier, Regier and Schultz (1997) suggested that branchiopods may be closer to other groups of arthropods than to malacostracan crustaceans, although there was no strong support for this arrangement and they concluded that the EF-1$\alpha$ data are ambiguous on this question. These authors later (1998b) depict remipedes closer to the crustacean stem, but again in this analysis, node support was not strong, and thus the authors remain suitably cautious as to interpretation of these data (J. Regier, pers. comm.). Spears and Abele (1997) conclude that "we cannot identify which crustacean lineage is most basal; branchiopods, pentastomes, branchiurans, and ostracodes [but note the absence of remipedes or cephalocarids] all diverged from the main crustacean lineage in relatively rapid succession." Although arguments on this point will surely continue for many years to come, we have elected to follow the 18S rDNA-based findings of Spears and Abele (1997), supported to some degree (in our estimation) by the EF-1$\alpha$ findings of Regier and Schultz (1997, 1998b; see also Shultz and Regier, 2000). Thus, we treat branchiopods first in our classification, thereby implying that we are in agreement with branchiopods being the most basal of the extant crustacean groups. This treatment also receives some support from Itô's (1989) suggestion of a remipede + cephalocarid + copepod clade, an arrangement that was also suggested by Spears and Abele (1997) based on 18S rDNA data (see especially their fig. 14.7 and accompanying discussion). We have not, however, created the additional taxonomic categories that would be required to group branchiopods as the sister group to all other crustaceans. In other words, our classification is far from being a strictly cladistically based arrangement. Branchiopods are thus accorded class status, as are the other five major crustacean groupings, in this classification. Additionally, if we are positing the branchiopods as the sister group to the other crustaceans, then we should list specific synapomorphies unique to the clade. Most of the morphological characters seeming to cast branchiopods in a primitive light (e.g., foliaceous limbs, anamorphic development) are indeed primitive features, but they may have been retained in this group and lost or modified in others. Noting simply that their morphology is "primitive" sheds no real light on phylogeny, and other groups of crustaceans exhibit other "primitive" characters. Possible candidates for branchiopod synapomorphies might include the "specialization of postnaupliar feeding apparatus to true filter feeding" (from Walossek, 1993:71), aspects of sperm morphology (Wingstrand, 1978), and the 18S rDNA sequences, which Spears and Abele (2000) used to conclude that "(1) branchiopods are monophyletic; (2) they are considerably divergent from other crustaceans (e.g., the Malacostraca), and (3) they are divided into two main lineages" (Anostraca and all others).

The issue of which extant class is closest to the

ancestral crustacean is of course not completely settled, and there are published arguments for presenting either the Cephalocarida or the Remipedia as the most primitive group of living crustaceans. There have also been, from time to time, hypotheses presented where other groups of crustaceans have occupied a basal position (e.g., McKenzie, 1991, postulated a bradoriid ostracode origin for all other crustaceans).

In favor of depicting remipedes as the most primitive class are the works of Schram (1986), Brusca and Brusca (1990), Briggs et al. (1993a), Schram and Hof (1998), Wills (1997), and Wills et al. (1998), all based on cladistic analyses of morphological characters from extant and extinct forms. Also supporting this view is the phylogeny presented by Jamieson (1991a) based on sperm ultrastructure, in which the Remipedia is the most basal of the crustacean groups. (It should be noted, however, that Jamieson's study is not purely independent of other phylogenies in that his figure is actually an overlay of the various sperm types on top of the classification offered by Schram in 1986.) Thus, there are workers at several independent laboratories whose studies have indicated that remipedes occupy the most basal position among the crustaceans, and several textbooks have followed this arrangement as well (e.g., Hickman et al., 1996: 401, figs. 20–30; Brusca and Brusca, 1990). Molecular evidence concerning where remipedes belong has been maddeningly difficult to obtain. Regier and Schultz (1998b) could not say with certainty (using EF-1$\alpha$), and Spears and Abele (1997) were equally unsure (using 18S rDNA). Emerson and Schram (1990, 1997) have also suggested that crustacean biramous limbs arose from fusion of adjacent uniramous limbs, and this has a bearing on the placement of remipedes relative to other crustacean groups as well (discussed further in Schram and Hof, 1998, but see Spears and Abele, 1997). It should also be pointed out that at least one publication (Moura and Christoffersen, 1996) suggests that the Remipedia are a derived assemblage that may be the sister group to the Tracheata (terrestrial mandibulates).

In support of cephalocarids occupying the most basal position among extant crustaceans are some surely primitive external morphological features. These features include the flattened and "Orsten-like" limbs, the lack of differentiation of the second maxilla (also shared with some of the Orsten crustaceans), and relatively anamorphic development. Hessler (1992) reviewed early considerations of the placement of the cephalocarids with respect to other crustaceans. He concluded, based on the morphology of some of the Upper Cambrian "Orsten" fauna of Sweden and in comparison with remipedes and other crustaceans, that the argument for placing cephalocarids at the base of the crustacean lineage is still strong (see also Walossek, 1993; Moura and Christoffersen, 1996). In Hessler's words, "among living crustaceans, cephalocarids still best personify what the ur-crustacean must have looked like." Hessler (1992) also made the point, with which we agree, that remipedes are quite specialized, and he found it "impossible to accept the claim that the Remipedia better approximates the ur-crustacean." However, cephalocarids face problems as primitive crustaceans as well. Schram and Hof (1998) point out some cephalocarid features they consider highly derived, and molecular studies (e.g., Spears and Abele, 1997; Regier and Schultz, 1998b) and spermatological data (especially lack of a flagellum; see Jamieson, 1991a) do not place cephalocarids basal to other crustacean taxa (although in fairness, the EF-1$\alpha$ data of Regier and Shultz do not decisively place cephalocarids elsewhere, either). We have not followed the suggestion of Hessler (1992) to revive the taxon Thoracopoda to include the cephalocarids, branchiopods, and malacostracans (based on their shared possession of an epipod on the trunk limbs).

### What Are the Relationships Among the Classes?

This question is closely related to the issues raised above. In fact, most of the competing phylogenetic hypotheses for class-level relationships have already been alluded to in earlier sections (e.g., in the sections "Cladistics and Classification of the Crustacea" and "Molecular Systematics and Classification of the Crustacea," and under the above three questions on crustacean monophyly, number of classes, and most primitive class). Rather than attempt a discussion of the many competing hypotheses for the relationships within and among the various classes, we have opted to treat each group individually below. We also refer the reader to the reviews by Wills et al. (1998) and Schram and Hof (1998), both in Edgecombe (editor, 1998, *Arthropod Fossils and Phylogeny*), and to the review of 18S rDNA studies by Spears and Abele (1997).

Concerning authorship of the name Crustacea, although most workers credit Pennant (1777), Lipke Holthuis, in a detailed and well-researched footnote to his FAO volume on marine lobsters (Holthuis, 1991), noted that the first usage was actually that of Brünnich in 1772. We have followed Holthuis' (1991) suggestion and have credited Brünnich (1772) with authorship of this taxon.

## CLASS BRANCHIOPODA

Virtually all evidence points to the fact that the branchiopods are a strongly supported monophyletic group, despite the staggering diversity of extant forms (e.g., see Martin, 1992). Lines of evidence indicating branchiopod monophyly include sperm morphology (Wingstrand, 1978), larval characters (e.g., Sanders, 1963), feeding apparatus (Walossek, 1993), adult characters (e.g., Negrea et al., 1999), and 18S rDNA sequence data (Spears and Abele, 1997, 1998, 1999a, b, 2000). However, the group's tremendous morphological diversity and age (see Fryer, 1987a–c, 1999; Martin, 1992;

Walossek, 1993; Negrea et al., 1999) makes it difficult to find characters shared by all extant members, and perhaps for this reason some analyses have hinted at para- or polyphyly (e.g., see Wilson, 1992). Gruner (1993) does not recognize the Branchiopoda, instead treating the extinct Lipostraca and the extant Anostraca and Phyllopoda (Notostraca + Diplostraca) as separate subclasses within the class Crustacea. The fact that there appears to be solid support from molecular data for branchiopod monophyly (e.g., Spears and Abele, 1997, 1998, 1999b, 2000) is nevertheless reassuring. There is also a consensus that, within the Branchiopoda, the Anostraca diverged early, are very primitive (despite a large number of apomorphic features in the various families), and should be depicted as separate from the remaining branchiopod groups. Beyond that, however, there is little agreement concerning the relationships among the constituent branchiopod taxa.

Because the Anostraca are clearly a separate lineage from the remaining branchiopods and are an ancient and slowly evolving group (e.g., see Fryer, 1992, 1999), we have elevated the group to the level of subclass, to be treated as the sister group of the other branchiopods (as was advocated also by Walossek, 1993, and Negrea et al., 1999). However, this move necessitates creating a name for the subclass or choosing an available name from the literature to contain the Anostraca (and which would eventually, we assume, contain also the fossil branchiopod order Lipostraca and possibly also the Cambrian *Rehbachiella*; see Walossek, 1993; Walossek and Müller, 1998). Tasch's (1969) proposal to use the name Sarsostraca (to contain anostracans and lipostracans) is not very appealing, in part because Tasch originally included in his Sarsostraca a noncrustacean (obviously also a nonbranchiopod), and one of his anostracans was in fact an insect larva (G. Fryer, pers. comm.). Nevertheless, the name Sarsostraca appears to be a valid preexisting name by ICZN standards and would have seniority over any newly proposed name here, so reluctantly we accommodate the order Anostraca within the subclass Sarsostraca, as did Bowman and Abele (1982) and, more recently, Negrea et al. (1999).

Finding a name suitable to contain the other (non-Anostraca) groups was more difficult. First of all, the tremendous morphological differences among the groups traditionally thought of as cladocerans, conchostracans, and notostracans has led several workers, most notable among them Geoffrey Fryer (e.g., see Fryer, 1987a, c, 1995, 1999a, b), to suggest that there is no reason to try to force such disparate groups into artificial groupings as "cladocerans" and "conchostracans." Fryer's well-written articles argue convincingly for the separation of these ancient and diverse taxa (most of which he would elevate to ordinal level), and indeed his suggested classification (Fryer, 1987a, c) has been followed by several workers, such as Martin (1992), Alonso (1996), Amoros (1996), Frey (1995), Brtek and Thiéry (1995), Thiéry (1996), Brtek (1997), and others. However, simply recognizing how different these groups are from one another and elevating the former conchostracan or cladoceran taxa to higher taxonomic categories while doing away with the categories that once included them does not, in our opinion, shed light on their relationships. The question still remains as to whether these orders are more closely related to one another than any is to some other crustacean assemblage. The morphological and molecular evidence seems to indicate (1) that branchiopods are monophyletic and (2) that some of these taxa (not all are well represented by molecular or even morphological data) are indeed related more closely to one another than to any other crustacean group. The alternative is to suggest that, for example, the Anomopoda are more closely related to anostracans or to some nonbranchiopod crustacean. We think this is very unlikely. Thus, the value of Fryer's arguments is in the recognition of the tremendous age and morphological differences that exist (and have existed for a long time) among these disparate taxa, a point that is well taken. Despite these arguments, and because we still must postulate relationships, we are forced to group these taxa together. Toward this end, several workers have suggested that we use the name Phyllopoda for the taxon encompassing the Notostraca and the bivalved branchiopods (see comments below about the nonmonophyly of the "diplostracans"), and indeed the name Phyllopoda has been used often for that assemblage (e.g., Walossek, 1993, and later). Unfortunately, the name Phyllopoda has also been used to denote groupings that include the Anostraca or that include the Ostracoda or that include the Leptostraca and Cephalocarida and in several other contexts as well. In fact, the term Phyllopoda has been used so often in crustacean systematics, and with such different meanings, that Martin and Christiansen (1995a) argued for avoiding it completely to avoid further confusion. Not surprisingly, we agree with Martin and Christiansen (1995a) and would prefer to employ another available name for this lineage. Does one exist? Tasch (1969) employed the names Calmanostraca (for the notostracans) and Diplostraca (for the conchostracans and cladocerans) as subclasses, but the two groups were treated equally (i.e., Tasch did not depict them as being more closely related to each other than either would be to the anostracans). Because the name Diplostraca obviously refers to the bivalved carapace seen in some groups, we could have opted to use the name Calmanostraca suggested by Tasch (1969) but expanding its definition to include both notostracans and the bivalved groups, which seems to be advocated by the classification proposed by Spears and Abele (2000). However, the name Calmanostraca should probably be reserved for containing the extinct Kazacharthra and the extant Notostraca (as it was first intended) when fossil taxa are eventually added to the "updated" classification (see also Negrea et al., 1999). Therefore, with trepidation and against our own recommendations (Martin and Christiansen, 1995a), we have resurrected the name Phyllopoda, using it this time to include the extant Notostraca and the bivalved branchiopod groups (i.e., all branchiopods except the Anostraca). We have credited the taxon name to Preuss (1951), who was, to our knowledge, the first person to use the name Phyllopoda in the sense that we are using it (to contain all branchiopods other than the anostracans). This decision will surely prompt arguments from many current students of the Branchiopoda (see especially Fryer, 1987c, 1995, 1999b).

There have been many significant findings in extant and extinct branchiopods that have altered our view of branchiopod relationships since the Bowman and Abele (1982) classification. Morphological treatments have included Fryer (1983, 1985, 1987a–c, 1995, 1996a, b, 1999), Martin (1992), Martin and Cash-Clark (1995), Walossek (1993, 1995), Olesen et al. (1997), Olesen (1996, 1998, 1999), Thiéry (1996), Amoros (1996), and Negrea et al. (1999), to mention only a few of the recent papers. There have also been several attempts to deduce branchiopod relationships using molecular data, including Hanner and Fugate (1997) and Spears and Abele (1997, 1998, 1999b, 2000). In the current classification, we have attempted to reconcile some of the recent morphological and molecular findings, but earlier classifications should not be discarded as being out of date or invalid. Indeed, many of the most detailed accounts of branchiopods remain the older, classical treatments, and to ignore these is a grave mistake. Thiéry (1996, based in large part on Martin, 1992) reviewed the biology of the noncladoceran groups (including *Cyclestheria* among the conchostracans), and Amoros (1996) reviewed the four "former cladoceran" orders Ctenopoda, Anomopoda, Onychopoda, and Haplopoda.

## SUBCLASS SARSOSTRACA, ORDER ANOSTRACA

Within the Anostraca, Brtek (1995) elevated the former chirocephalid subfamily Artemiopsinae to family level and thus recognized the Artemiopsidae. Earlier, Brtek (1964) established the family Linderiellidae. However, Denton Belk (pers. comm.) believed these moves are unwarranted. Concerning the Artemiopsidae, Belk stated, "placing this single genus in a separate family obscures the many features it shares with other genera in the Chirocephalidae, and is thus a hindrance to having a meaningful taxonomic classification of the Anostraca." Concerning the Linderiellidae, he noted that "these genera have antennal appendages and some penal features that suggest they are related to other genera of the Chirocephalidae; separate familial status obscures these seemingly significant similarities." In light of Belk's expertise with anostracans, we have followed his suggestion and have not recognized these two families, although they are recognized in the latest key to families and genera (Brtek and Mura, 2000). Our classification of the Anostraca therefore follows Belk (1996), with the exception of the Linderiellidae (which was included by Belk, 1996, but is not included here). A recent molecular analysis (Remigio and Hebert, 2000) of the relationships among extant anostracan families suggested two clades, one containing Artemiidae and Branchipodidae and the other containing the other five families.

## SUBCLASS PHYLLOPODA

By placing anostracans in a subclass separate from all other branchiopods, we are assuming also that the other branchiopods form a monophyletic grouping. In other words, we believe that the notostracans, conchostracans, and cladocerans are more closely related to one another than any of those groups is to the anostracans. There are some morphological features (e.g., Negrea et al., 1999) and molecular data (e.g., Spears and Abele 1997, 1999b, 2000) that suggest this might be true. This arrangement has been proposed by many other workers as well (some of whom, such as Walossek, 1993, 1995; Walossek and Müller, 1998, have also employed the name Phyllopoda in the same sense that we are using it).

## ORDER NOTOSTRACA

It may be necessary, once fossil taxa are included in this classification, to someday resurrect Tasch's (1969) name Calmanostraca to accommodate the extant notostracans and the extinct and obviously closely related Kazacharthra. The sole family of extant Notostraca, Triopsidae, is credited to Keilhack ("Kielhack" was a misspelling in Bowman and Abele, 1982), and that date has been changed from 1910 to 1909 (L. Holthuis, pers. comm.). Although the original spelling was Triopidae, as listed in Bowman and Abele (1982), the spelling Triopsidae (based on the genus *Triops*) was entered in the Official List of Family-Group Names in Zoology by the ICZN, Opinion 502 (M. Grygier, pers. comm.).

## ORDER DIPLOSTRACA

As noted above, the Phyllopoda as used here includes the orders Notostraca and Diplostraca (a name that predates Onychura used by some authors, such as Walossek, 1993, and Negrea et al., 1999). Whether these are indeed sister taxa is unclear; there is some morphological and molecular evidence to suggest that this might not be the case. Recognition of the taxon Diplostraca indicates our feeling that the former conchostracan and cladoceran groups are indeed related. There appears to be some morphological (e.g., see Walossek, 1993; Olesen, 1998; Negrea et al., 1999) and molecular (Spears and Abele, 2000) evidence supporting this relationship, although the view is certainly not universally shared (e.g., see the exchange between Olesen, 1998, 2000, and Fryer, 1999, 2001), and there is a large body of evidence suggesting that Diplostraca is nonmonophyletic. Additionally, there is considerable doubt concerning the monophyly of some of the groups we have included within it, such as the Cladocera. Fryer (1987a, 1995, 1999a, b) discusses the great morphological differences among the four groups traditionally placed in the "so-called Cladocera" and highlights the trenchant differences among these taxa and the difficulty in reconciling these forms within one taxonomic category. We should also point out that the "secondary shield" mentioned as unifying these taxa (e.g., by Walossek, 1993; Olesen et al., 1997; Olesen, 1998) is, according to Fryer (1996b, 1999b), simply nonexistant, a misunderstanding of the nature of the crustacean carapace. Other characters that supposedly unite the "diplostracan" groups are similarly called into question by Fryer in a series of papers (1987a–c, 1995, 1996a, b, 1999b). In particular, after considerable work in attempting to reconstruct a primitive anomopod from which extant anomopods could have been derived and by so doing highlighting the great difficulties of any such exercise, Fryer (1995) argued against attempting to force such disparate taxa as *Leptodora, Bythotrephes,* and the superficially similar ctenopods into a taxon with the Anomopoda, stating (pers. comm.) that "when those who make these proposals can support them by evolutionary series that involve animals that would work, I'll pay more attention to them."

Within the Diplostraca, we have removed the "Conchostraca" (following to some extent the suggestions of Fryer, 1987c, and Olesen, 1998) in recognition of (1) the distinct nature of the Laevicaudata (Lynceidae), (2) the stark differences that separate *Cyclestheria hislopi* (sole member of the Cyclestheriidae) from all other conchostracans, and (3) *Cyclestheria*'s possible affinities to the cladocerans on morphological and molecular grounds (see Martin and Cash-Clark, 1995; Olesen et al., 1997; Olesen, 1998; Spears and Abele, 1998, 2000). The fact that *Cyclestheria* differs significantly from other spinicaudate conchostracans, and probably to the extent that it should not be placed among them, has also been highlighted (Martin and Cash-Clark, 1995; Olesen et al., 1997; Olesen, 1999; Negrea et al., 1999). Thus, our resulting classification within the Diplostraca differs slightly from, and is in some ways a compromise between, the classification suggested by Olesen (1998) based on morphological characters and that suggested by Spears and Abele (2000) based on molecular data and is easily reconciled with the phylogeny proposed by Negrea et al. (1999). Our arrangement does not agree with the somewhat preliminary findings of Hanner and Fugate (1997) based on a relatively small segment of the genome.

Removal of *Cyclestheria* from the Spinicaudata and placing it on an equal footing with the remaining Spinicaudata and with the Cladocera necessitated the creation of a separate suborder, the Cyclestherida, which we are crediting to Sars (1899) in keeping with ICZN article 50.3.1. Negrea et al. (1999) used the same spelling to refer to an order (Cyclestherida) within their superorder Conchostraca, thus indicating a closer affinity of *Cyclestheria* to the conchostracans rather than the cladocerans. We have not taken the bolder step of actually including the Cyclestheriidae among the Cladocera, although there is apparently evidence for this as well. Spears and Abele (1999a, b, 2000) note that, not only do 18S rDNA sequence data support the close relationships of *Cyclestheria* and the cladocerans, the two groups also share certain hypervariable regions of the gene that are not found in other branchiopods, and these are potential synapomorphies. Ax (1999) first suggested the term "Cladoceromorpha" for the clade containing *Cyclestheria* plus Cladocera. Papers by Crease and Taylor (1998) and Taylor et al. (1999) appear to offer additional molecular support, and the phylogeny suggested by Negrea et al. (1999:196) supports such a clade as well, although their resulting classification of the Branchiopoda into five superorders does not.

Sassaman (1995) presented fascinating insights into possible phylogenetic models for the conchostracan families based on the evolution of unisexuality in the group; he views lynceids as the sister group to all other families, while noting at the same time the unusual nature of the cyclestheriids, which he posits as the sister group to the remaining "spinicaudatan" families. Thus, in many ways, Sassaman's (1995) phylogeny is consistent with our classification.

Within the former "conchostracan" groups, the spelling of the Lynceidae has been corrected (from Lyncaeidae, a typographical error in Bowman and Abele, 1982), and authorship for the family is now credited to Baird, 1845 (L. Holthuis, pers. comm.). Mark Grygier points out (pers. comm.) that ICZN Opinion 532 attributes the family name to Sayce, 1902; however, there are clearly earlier uses of the family name Lynceidae (e.g., see review by Martin and Belk, 1988), and we are crediting the family name to Baird as noted above.

Although the genera *Imnadia* and *Metalimnadia* at times have been suggested to represent distinct families (the Imnadiidae Botnariuc and Orghidan and the Metalimnadiidae Straskraba; see Marincek and Petrov, 1991; Roessler, 1991, 1995a, b; Orr and Briggs, 1999:8), most workers (e.g., Martin, 1992; Sassaman, 1995) consider them members of the family Limnadiidae, as do we. Roessler's (1991) erection of the family Paraimnadiidae was based on a species he described as *Paraimnadia guayanensis,* a junior synonym of *Metalimnadia serratura* (see Orr and Briggs, 1999). We also include among the limnadiids the genus *Limnadopsis* and agree with

Bowman and Abele in not recognizing Tasch's (1969) family Limnadopsidae.

The superfamilies Cyzicoidea (which contained only Cyzicidae) and Limnadioidea have been removed, as there is no longer any need for them in light of the above reassignments. Indeed, the families Cyzicidae and Leptestheriidae are probably more closely related to each other than either is to the Limnadiidae (Martin, 1992; Sassaman, 1995).

Within the Cladocera, the spelling of the Holopediidae has been corrected (from Holopedidae in Bowman and Abele, 1982) in light of the spelling of the type genus *Holopedium* (M. Grygier, pers. comm.). The correct spelling of Macrotrichidae (rather than Macrothricidae) was also pointed out to us by M. Grygier (pers. comm), referring us to Appendix D of the ICZN, third edition, example 24, page 223 (ICZN, 1985a), for examples of family names formed from genus names ending in -*thrix*. However, the fourth edition of the Code (ICZN, 1999) now allows such misspellings to stand if they are in "prevailing use," which the family name Macrothricidae certainly is. Thus, we retain the spelling Macrothricidae. (This same logic (i.e., retention of a misspelling because of prevailing use) applies also to the family Rhizothricidae in the harpacticoid copepods.)

Within the Anomopoda, we have removed the family Moinidae, following the suggestion of G. Fryer (1995, and pers. comm.). Comparisons of the trunk limbs of species of *Moina* and *Daphnia* indicate great similarity between these groups; certainly they are much more similar than are many macrothricid and chydorid genera to each other. If a separate family were recognized for *Moina* and *Moinodaphnia,* then we would have to erect a series of families for various chydorids and macrothricids, which we see as only adding to the confusion. Thus, the Moinidae is not recognized here. For the same reason, we have decided not to recognize the family Ilyocryptidae as treated by Smirnov (1992) based on the genus *Ilyocryptus* (see also Young, 1998:23). However, it is possible that the correct course of action would be to acknowledge anomopodan diversity by recognizing both the Moinidae and Ilyocryptidae as valid families and establishing the additional families for other genera as needed.

The four main cladoceran groupings have been treated as infraorders. Although we are in full agreement with Fryer's (1987a–c, 1995) assessment of the distinct nature of, and tremendous differences among, these taxa (Fryer argued for removal of the terms "cladocera" and "conchostraca" as formal taxonomic entities), we nevertheless felt that the four groups are more closely related to one another than any one of them is to any other crustacean assemblage, the same conclusion reached by Richter et al. (2001) and several earlier workers. This may prove to be a mistake. Certainly, treatment of the cladocerans as a single order containing four infraorders and a handful of families has the unfortunate appearance of minimizing the staggering morphological and ecological diversity of this group, and we very much regret that. Schwenk et al. (1998) provided a preliminary estimate of the relationships of the Ctenopoda, Haplopoda, Onychopoda, and Anomopoda based on 16S rDNA sequence data. See Fryer (1995) for suggested relationships among the families of the Anomopoda and Richter et al. (2001) for 12S rDNA-based relationships among onychopods and between the "gymnomerans" (= onychopods + *Leptodora*) and other cladoceran groups.

The taxon "Eucladocera" has been removed, as we saw no evidence for grouping together all other cladocerans as the sister taxon to the monotypic Haplopoda (*Leptodora*), as proposed by several workers (most recently by Negrea et al., 1999). Our classification is more in keeping with the study by Richter et al. (2001), who supported the monophyly of the Onychopoda + Haplopoda (the former Gymnomera) and argued for cladoceran monophyly. The superfamilies Sidoidea, Daphnioidea, and Polyphemoidea have also been removed.

## CLASS REMIPEDIA

It is a little discouraging that we still know so little about the phylogenetic relationships of this fascinating group. The initial establishment of a separate class (Yager, 1981) met with criticism early on, and similarities between the limbs of remipedes and those of certain maxillopods have been pointed out (Itô, 1989). Felgenhauer et al. (1992) hinted at molecular data that suggested maxillopodan affinities as well, although, to our knowledge, these data have not been published. Spears and Abele (1997) also suggested possible maxillopodan affinities. In an early draft of this classification, we had the remipede families included among the Maxillopoda, but this was criticized, and rightly so, by several persons who pointed out that some of the similarities between Remipedia and Maxillopoda are symplesiomorphies (although others, such as the loss of the maxillary endopod, defined precoxa of the maxillule, and three-segmented endopod of the trunk limbs, may be synapomorphies) and are insufficient to warrant the inclusion of the former among the latter. More detailed morphological studies (e.g., Schram et al., 1986; Itô and Schram, 1988; Schram and Lewis, 1989; Yager, 1989a, b, 1991; Yager and Schram, 1986; Emerson and Schram, 1991; Felgenhauer et al., 1992) seem to confirm the unique nature of the group. Their status as a distinct class is therefore maintained in this classification. See also our introductory comments concerning which class of extant Crustacea appears most plesiomorphic

As noted above in the general discussion of the primitive groups of Crustacea, several workers (e.g., see Schram, 1986; Brusca and Brusca, 1990; Briggs et al., 1993a; Schram and Hof, 1998; Wills, 1997; Wills et al., 1998) have suggested that remipedes occupy the most basal position among the extant crustaceans. These arguments are perhaps best summarized in Schram and Hof (1998) and in Wills (1997), where remipedes come out at the base of all other Crustacea groups following cladistic analyses of large datasets. Moura and Christoffersen (1996) take an opposing stance, suggesting that remipedes are an apical group of crustaceans that are possibly the sister group to terrestrial mandibulates. To us, the evidence (morphological, molecular, and developmental) for branchiopods being basal appears stronger (see earlier comments on primitive crustaceans). Emerson and Schram (1990; see also Emerson and Schram, 1991) have suggested that crustacean biramous limbs may have arisen from fusion of adjacent uniramous limbs, and this has a bearing on the placement of remipedes relative to other crustacean groups (discussed further in Schram and Hof, 1998). Spears and Abele (1997) also discussed possible affinities between remipedes and cephalocarids, some of which may be artifactual as a result of long branch attractions.

Within the Remipedia, the order Nectiopoda was erected by Schram (1986) to separate extant remipede families from some fossils that appear remipedian (and that are treated as the fossil order Enantiopoda). One additional family, the Godzilliidae, was added by Schram et al. (1986). Yager and Humphreys (1996) reported the first species from Australia and the Indian Ocean and presented a key to the world species known at that time. Cals (1996) reviewed the biology of the group and presented a table comparing the characteristics of the two currently accepted families, Speleonectidae and Godzilliidae; more recently, Yager and Carpenter (1999) and Carpenter (1999) have added to what is known of the natural history of speleonectids.

## CLASS CEPHALOCARIDA

Our classification differs from that of Bowman and Abele (1982) only in recognizing a single family, Hutchinsoniellidae, rather than two families. The family Lightiellidae proposed by Jones (1961) is thought to differ only slightly and insignificantly from the characters established for the former family (R. Hessler, pers. comm.). Our placement of the cephalocarids here, between the remipedes and maxillopods, to some degree reflects the summary finding of Spears and Abele (1997) that remipedes and cephalocarids may constitute a clade that is the sister group to one of the maxillopodan groups (the Copepoda) (e.g., Spears and Abele, 1997, figs. 14.4, 14.7, and accompanying text), although Spears and Abele (1997) also note that this arrangement is not well supported by their bootstrap analysis. The placement of cephalocarids and remipedes together, and adjacent to the maxillopods, in some ways also supports Itô's (1989) morphology-based suggestion of a remipede + cephalocarid + copepod clade. Hessler and Elofsson (1996) recently reviewed what is known of cephalocarid biology and phylogeny.

## CLASS MAXILLOPODA

The Maxillopoda continues to be a terribly controversial assemblage concerning both the number of constituent groups and the monophyly of the entire taxon. We were tempted to abandon, once and for all, the concept of a monophyletic Maxillopoda, as there seems very little in the way of morphological or molecular evidence uniting the disparate groups (Wilson, 1992; Spears and Abele, 1997; Shultz and Regier, 2000). Ostracodes in particular have been placed sometimes within the Maxillopoda (e.g., see Boxshall and Huys, 1989a) and sometimes in their own class, and the issue remains unresolved despite much debate (e.g., see Boxshall et al., editors, *Acta Zoologica,* vol. 73(5), 1992). It is certainly no secret that the characters used in defining the group do not hold for many of the taxa traditionally thought of as being "maxillopodan." Abandoning the Maxillopoda seems to have been implied in tome VII fascicule II of the *Traité de Zoologie* (1996), as only the constituent groups are treated with no mention of maxillopod affinities or relationships (e.g., see Grygier 1996a, b), and Gruner (1993) similarly did not recognize the Maxillopoda. Boxshall (1983) and others have argued against recognition of the Maxillopoda on morphological grounds, although Boxshall has also continued to employ it from time to time (e.g., in Huys et al., 1994). Yet other workers (e.g., see Newman, 1983; Grygier, 1983a; Walossek, 1993; Wills, 1997; Walossek and Müller, 1998) have argued, some quite forcefully, that there is merit to recognition of the Maxillopoda as a natural (monophyletic) assemblage, despite the fact that there seem to be exceptions to every synapomorphy proposed. In fairness, so many maxillopodan taxa are so small and/or modified as parasites that it should come as no surprise to find exceptions to groundplans. Removal of the Maxillopoda as a class would raise the number of crustacean classes from six to nine once the maxillopodan subclasses were elevated (each to the level of class).

The somewhat controversial history of the concept of the Maxillopoda (whether it is monophyletic, and if so, which groups should be included, and what the relationships are within the group and of the group to other crustaceans) is reviewed on morphological grounds by Grygier (1983a, b, 1985, 1987a–c), Müller and Walossek (1988), Boxshall and Huys (1989a), Huys (1991), Newman (1992), Schram et al. (1997), Schram and Hof (1998), and papers cited therein, and on molecular grounds by Abele et al. (1992), Spears et al. (1994), and Spears and Abele (1997). Some of the fossil discoveries since the Bowman and Abele classification have a bearing on our understanding of the monophyly and definitions of the Maxillopoda as well, such as the description of the Skaracarida

(Müller and Walossek, 1985), the Orstenocarida (Müller and Walossek, 1988), and the Mazon Creek Cycloidea (Schram et al., 1997). A relatively recent and widely used text on invertebrates (Brusca and Brusca, 1990) recognizes the Maxillopoda (including the Ostracoda), and that text is often cited in other listings of crustaceans (e.g., the Tree of Life web project; see URL http://ag.arizona.edu/tree/eukaryotes/animals/arthropoda/crustacea/maxillopoda.html), whereas another recent text (Gruner, 1993) treats the various maxillopod groups separately.

While it is clear that there is not a single "good" character shared by the various maxillopod groups (see especially Boxshall, 1992), it is also true that some of them seem closely related on morphological and molecular grounds. Furthermore, even some of the more vocal opponents to the Maxillopoda will argue from time to time that there seems to be a core group of taxa that "hang together well" (although the members of this core group change depending on the speaker). The question as to which groups are and which are not "true" maxillopods and whether any of the constituent groups should remain allied in a classification has not been, in our opinion, satisfactorily answered.

Although the issue is still unresolved, we have found it useful to continue to recognize the Maxillopoda, and refer the reader to discussions of morphological characters seeming to unite the maxillopodan groups (see above). At the same time, we caution readers that acceptance of the Maxillopoda as monophyletic and acceptance of the constituent groups are not universal and nowhere near as finalized as envisioned by Walossek (1993; see review of this work by Martin, 1995) or by Walossek and Müller (1994). In the latter paper, Walossek and Müller state that the "interrelationships of the majority of maxillopod taxa, particularly of the thecostracan lineage, are well-founded on morphological, ontogenetic, and fossil data." This could hardly be further from the truth. We have followed, for the most part, the treatment by Newman (1992) for higher classification of the Maxillopoda and his subsequent work (especially Newman, 1996) for lower taxonomic divisions. We differ from Newman's treatment in not using the "superclass" rank, in an attempt to be consistent with our other uses and categories. This necessitated the creation of some lower level taxonomic names (superorders, infraorders, etc.) that unfortunately add to the clutter of this already confusing assemblage. We also differ from Newman's treatment in that we have treated the Rhizocephala as members of the cirripedian line (see below), as suggested by J. Høeg (pers. comm.) and others (see below).

Published and unpublished hypotheses of relationships within the Maxillopoda are numerous. As one example, Walossek and Müller (1998) feel that there are two rather clear lines and presented character states for each. The first is the "copepod line," including the copepods, mystacocarids, and the extinct Skaracarida (which is in keeping with the analysis of maxillopod orders by Boxshall and Huys, 1989a). The second is the "thecostracan line" that includes the tantulocarids, ascothoracidans, facetotectans, acrothoracicans, and cirripeds. However, this division does not appear to have much neontological (e.g., Høeg, 1992a) or molecular (Spears et al., 1994; Spears and Abele, 1997) support. Some of the major areas of disagreement in the various maxillopod hypotheses include whether the ostracodes should be included vs. excluded, where the Facetotecta belong, where the Tantulocarida belong, and the placement (and subdivision) of the cirripedes. We have attempted to list the more salient of these efforts in the individual sections that follow. For an overview of maxillopod classification and phylogenetic studies, we refer readers to Grygier (1987a, b), Newman (1987), Boxshall and Huys (1989a), Boxshall (1992), Huys et al. (1993), Spears et al. (1994), and Spears and Abele (1997).

## SUBCLASS THECOSTRACA

Spears et al. (1994) concluded, based on 18S rDNA sequence data, that the Thecostraca, as recognized by Grygier (1987a; see also Grygier, 1987b) and Newman (1987, 1992) on morphological grounds, is a monophyletic assemblage. Furthermore, within the Thecostraca, Spears et al. (1994) recognized two major subdivisions, one containing the Ascothoracida and a second (a modified "Cirripedia") containing the Acrothoracica, Rhizocephala, and Thoracica. Although we have maintained the Thecostraca, we have not divided the group as suggested by Spears et al., treating instead the Facetotecta (which was not treated by Spears et al.), Ascothoracida, and Cirripedia (now including the Acrothoracica, Rhizocephala, and Thoracica) as taxa of equivalent rank (infraclasses in the current scheme) within the Thecostraca. Huys et al. (1993) recognized the Thecostraca (without the tantulocarids) and postulated a sister-group relationship between the Tantulocarida and Thecostraca, noting that "inclusion of the Tantulocarida in the Thecostraca, as proposed by Newman (1992), would significantly dilute the otherwise robust concept of the Thecostraca." Jensen et al. (1994b) described cuticular autapomorphies (details of the lattice organs; see also Høeg et al., 1998) that also support the Thecostraca as a monophyletic assemblage.

## INFRACLASS FACETOTECTA

Surely one of the biggest remaining mysteries of crustacean classification is the taxon Facetotecta. Credited to Grygier (1985, corrected from 1984 in Bowman and Abele by M. Grygier, pers. comm.; see also Grygier, 1987a, b, 1996a), the taxon currently contains no further taxonomic divisions other than a single genus, *Hansenocaris* Itô, to accommodate the curious "y-larvae." The group consists of small (250–620 micrometers) nauplii with a vaulted and ornamented cephalic shield, sometimes with complex honeycomb patterns, followed by a relatively long and ornamented trunk region. The intriguing possibility that these planktonic forms may be larval tantulocaridans (which would result in tantulocaridans being classified under the Facetotecta) has also been suggested (M. Grygier and W. Newman, pers. comm.), based in part on the fact that there are still gaps in the known life cycle of tantulocarids following the work of Boxshall and Lincoln (1987) and Huys et al. (1993). As Grygier (pers. comm.) points out, "there is a hole in the tantulocaridan life cycle where y-larvae might fit (i.e., the progeny of the supposedly sexual males and females), but it would be a very tough fit." Newman (pers. comm.) succinctly describes the current state of our knowledge: "They [facetotectans] are the larvae of some very small, parasitic maxillopodan, and if not tantulocarids, they are the last survivors of some other great free-living radiation close to them." A recent review of the Facetotecta was provided by Grygier (1996a).

## INFRACLASS ASCOTHORACIDA

The Ascothoracida have been treated in the past sometimes as an order (e.g., by Newman, 1992), but that rank is changed to infraclass here to accommodate the constituent taxa that have been elevated to (or treated as) orders by Grygier (1987a, b) and Newman (1987, 1996), whose classifications we follow (see also Grygier, 1983a, b, 1987c, 1996b). Our classification thus includes two families, Ascothoracidae Grygier, 1987, and Ctenosculidae Thiele, 1925, that were not included in the Bowman and Abele (1982) listing. Thus, the infraclass currently consists of two orders, Laurida and Dendrogastrida, each with three families.

## INFRACLASS CIRRIPEDIA

Whether the Cirripedia should include the Rhizocephala (e.g., Høeg, 1992a) or whether the Rhizocephala are early offshoots of the cirripedian line and not members of the crown group (as in Newman, 1982, 1987; Grygier, 1983a; Schram, 1986) is not settled. However, there appears to be a growing consensus that the Rhizocephala and the Cirripedia form a monophyletic group. Høeg (1992a) provides strong evidence based on larval morphology, and Spears et al. (1994) support this with molecular data. There is some evidence (both morphological and molecular) that Cirripedia, with or without the Rhizocephala, may be paraphyletic (Newman, 1987; Spears et al., 1994). Our classification treats the Cirripedia as one of three infraclasses of the subclass Thecostraca. Included in our Cirripedia are the Rhizocephala. This is more in line with Høeg's (1992a) view, where he suggested that Cirripedia be defined as containing the Rhizocephala, Thoracica, and Acrothoracica, than with Newman's (1992) view, although Newman (pers. comm.) has indicated to us more recently that he now agrees with placing the rhizocephalans within the Cirripedia. Characters of the naupliar and cypris larval stages argue for inclusion of the rhizocephalans within the Cirripedia (Høeg, 1992a), and molecular evidence (in the form of rRNA sequences) supports this (Spears et al., 1994). A close relationship between Rhizocephala and Thoracica is supported by 18S rDNA data as well (Abele and Spears, 1997).

Although earlier molecular studies (Spears et al., 1994) seemed to indicate that the Ascothoracida might be the sister taxon to the Acrothoracica (which we have included in the Cirripedia), further analyses have not supported this arrangement (Spears and Abele, 1997). Thus, our current arrangement maintains the inclusion of the Acrothoracica within the Cirripedia.

Treatment of the Iblomorpha as one of four thoracican suborders (with no phylogenetic order implied) is at least in keeping with the finding of Mizrahi et al. (1998) that *Ibla* is not as different from other thoracicans as some earlier workers had supposed and should not be treated as near the base of the stem of the Thoracica.

An extensive morphology-based cladistic analysis of the Cirripedia Thoracica by Glenner et al. (1995), reanalyzed with some characters rescored by Høeg et al. (1999), supported the monophyly of the Balanomorpha and Verrucomorpha and suggested that several groups, among them the Pedunculata, Scalpellomorpha, and Chthamaloidea, were demonstrably paraphyletic. Yet other major questions remained unresolved, and Glenner et al. (1995) suggested that the fields of larval ultrastructure, early ontogeny, and molecular sequencing might be promising areas for future research. Anderson (1994:326) presented a slightly different classification, where the Cirripedia (which he treats as a subclass within the class Thecostraca) comprises five superorders (two of which, the Archithoracica and Prothoracica, would be new taxa coined by him), but this has not been followed by many other workers. Naupliar evidence seems to support, in general, the classification we have depicted within the cirripedes based on adult morphology (Korn, 1995). Høeg (1995) presents some interesting alternatives based on evolution of the sexual system of cirripedes and related groups, where again thecostracans and tantulocaridans are depicted as sister taxa.

A study of the brachylepadomorphs (Newman, 1987) led Newman to abandon thoughts of polyphyly in favor of monophyly of the sessile barnacles (Newman, 1991, 1993, 1996, and pers. comm.). Thus, the Sessilia was resurrected to contain the brachylepadomorphs, verrucomorphs, and balanomorphs, as was the Penduculata for the pedunculate barnacles. This has been challenged by Glenner et al. (1995) (see above and see also the reanalysis of the Glenner at al. data by Høeg et al., 1999).

A review of various bodies of information concerning barnacle evolution (Schram and Høeg, 1995) reveals mostly that we still have much to learn about the relationships of the various groups of maxillopods.

## SUPERORDER ACROTHORACICA

For this group, we have followed the classification of Newman (1996), where acrothoracicans are divided among two orders, Pygophora (with two families) and Apygophora (with a single family).

## SUPERORDER RHIZOCEPHALA

Our classification of this group follows Høeg (1992), Høeg and Rybakov (1992), Høeg and Lützen (1993, 1996), Huys (1991), and Lützen and Takahashi (1996). Thus, we treat the Rhizocephala as an infraclass that contains two orders, Kentrogonida (with three families) and Akentrogonida (with six families), although there is concern that one or both of these orders may be paraphyletic (see Høeg and Lützen, 1993). Jensen et al. (1994a, b) supported monophyly of the Akentrogonida on the basis of details of the lattice organs.

Within the Kentrogonida, concerning the issue of authorship of the families Peltogastridae and Sacculinidae (which we had earlier credited to Boschma), W. Vervoort writes (pers. comm.): ". . . both the families Peltogastridae and Sacculinidae must be ascribed to Lilljeborg, 1860. This has been duly checked. Boschma lived [from] 1893–1976 and cannot possibly be the author of these two families. Holthuis and I consulted Lilljeborg's 1860 publication, a copy of which is in our library; there is not a shadow of a doubt concerning his authorship." The family Sylonidae (Sylidae in Bowman and Abele) has been subsumed within the Clistosaccidae Boschma, which is now included in the Akentrogonida (J. Høeg, pers. comm.).

Within the Akentrogonida, three new families (Duplorbidae, Mycetomorphidae, and Thompsoniidae) were described by Høeg and Rybakov (1992) and one new family (Polysaccidae) was added by Lützen and Takahashi (1996). The Chthamalophilidae is recognized as a valid family (also following Høeg and Rybakov, 1992), and, as noted above, the Clistosaccidae was transferred into the Akentrogonida from the Kentrogonida.

## SUPERORDER THORACICA

Although few new extant families have been suggested since 1982, there have been significant rearrangements of the cirripedes (or attempts to rearrange them) by workers using morphological and molecular data. Perhaps the most comprehensive is the cladistic study by Glenner et al. (1995), who concluded that many currently recognized groups appear to be paraphyletic, including the groups that appear in our classification under the headings "Lepadomorpha" and "Pedunculata." However, Glenner et al. (1995) also noted that "we have far to go before a new taxonomy can emerge" and suggested the continued use of such commonly used terms as "lepadomorphs" or "pedunculates" as long as workers understand that these are groupings more of convenience than of common descent. We are not in agreement with this philosophy and would prefer to recognize taxa that reflect common descent, but in this group, it is apparent that we are not yet at the point where we know which clades are valid.

For the most part, we have followed the classification of the Thoracica given by Newman (1996). Thus, we are recognizing the order Pedunculata (an old name that was previously thought to lack validity but that Newman (1996) feels is a natural assemblage and thus has resurrected) as containing four suborders. Some of the names in this order (e.g., Heteralepadomorpha, Iblomorpha, Scalpellomorpha) are credited to Newman (1987), although it is clear that these higher taxon names are based on older works, which perhaps should be credited as the taxon author and date if we were to closely adhere to ICZN article 50.3.1 as extended to higher taxa. Many of the families now treated in these four suborders were elevated from subfamily status by Newman (1987). For example, within the Scalpellomorpha, only the family Scalpellidae Pilsbry is also found in the Bowman and Abele (1982) classification. Within the resurrected order Sessilia (see Newman, 1987; Buckeridge, 1995), the brachylepadomorph family Neobrachylepadidae was described by Newman and Yamaguchi (1995) and the verrucomorph family Neoverrucidae was described by Newman (1989, in Newman and Hessler, 1989:268; see also Newman, 1989). Within the Balanomorpha, Buckeridge (1983) added the superfamily Chionelasmatoidea, containing the single family Chionelasmatidae. Suggestions for evolutionary radiations within the Balanomorpha were presented by Yamaguchi and Newman (1990). A recent molecular analysis of several thoracican taxa (Harris et al., 2000) suggests that the sessile barnacles are monophyletic but that the pedunculate forms (our Pedunculata) may not be.

## SUBCLASS TANTULOCARIDA

The Tantulocarida, bizarre parasites of other deep-sea crustaceans, were known as early as the beginning of the 20th century (reviewed by Huys, 1990e, 1991; Boxshall, 1991, 1996) but were recognized as a distinct class of Crustacea only in 1983 (Boxshall and Lincoln, 1983), just too late for inclusion by Bowman and Abele (1982). They have since been relegated to a subclass or infraclass within the Thecostraca or have been proposed as the sister group to the Thecostraca within the Maxillopoda (e.g., Boxshall and Huys, 1989a; Boxshall, 1991; Huys et al., 1993). Our classification follows that of Huys (1990e) (see also Huys, 1991, where two families are also described). Discussions of the relationships of tantulocaridans (all of which lack recognizable cephalic limbs, other than paired antennules in one known stage, which makes elucidation of their affinities very difficult) to other Crustacea can be found in the above works as well as in Boxshall and Lincoln (1987) and Huys et al. (1993). Newman (pers. comm.) feels that, based on the placement of the male and female genital apertures and based also on the fact that the male genital aperture empties at the end of a median intromittant organ, tantulocarids are so closely related to the Thecostraca that placement *within* the subclass Thecostraca may be warranted. Certainly they appear more closely related to the Thecostraca than to any other maxillopodan group (W. Newman, pers. comm.; J. Høeg, pers. comm.; and some of the above references). However, for the present classification, we have retained them as a separate group within the Maxillopoda but not within the Thecostraca. Separate status of the Thecostraca and Tantulocarida was also suggested on morphological grounds by Boxshall and Huys (1989a), although we have not closely followed their proposed arrangement (their fig. 6) for the organization of the Maxillopoda.

The unusual and confusing life cycle of the tantulocarids is now more completely known, thanks to the work of Boxshall and Lincoln (1987) and Huys et al. (1993). Based on these works and because of the gap still remaining in the known tantulocarid life cycle, the possibility that y-larvae (the Facetotecta) might belong to this taxon has at least been considered (M. Grygier, pers. comm.; see earlier discussion under Facetotecta).

## SUBCLASS BRANCHIURA

To our knowledge, this subclass, containing a single order and family, has not changed since Bowman and Abele (1982) (see also Gruner, 1996). Bill Poly (pers. comm.) alerted us to the fact that, although the order Arguloida is often credited to Rafinesque (1815), Rafinesque employed only the term "Argulia" without treating it as a family or order. The first person to use the name as an order was apparently S. Yamaguti (1963, as Argulidea) (B. Poly, pers. comm.). Bowman and Abele (1982) credited the family name to Leach (1819) (as did Yamaguti, 1963, and Gruner, 1996). Although Leach's usage appeared after Rafinesque's work, we have credited Leach with recognition of the family and Yamaguti (1963) for the order, despite Rafinesque's original (1815) use of "Argulia," which of course became the basis of both family and order names. Yamaguti (1963) also established the family Dipteropeltidae, and some subsequent workers (e.g., Overstreet et al., 1992; Young, 1998) have continued to recognize it, although we do not.

## SUBCLASS PENTASTOMIDA

One of the most contentious changes in the new classification is the inclusion within the Crustacea Maxillopoda of the former phylum Pentastomida, all members of which are, as adults, parasites in the respiratory passages of vertebrates (see reviews by Riley, 1986, and Palmer et al., 1993). An alliance between pentastomes and branchiuran crustaceans was first suggested on the basis of sperm morphology some 29 years ago (Wingstrand, 1972; see also Wingstrand, 1978; Riley et al., 1978; Grygier, 1983). Inclusion of pentastomes among the Crustacea was actually considered but rejected by Bowman and Abele (1982), who at the time felt that insufficient evidence was available on that issue. Ironically, it was Abele et al. (1989) (see also Abele et al., 1992) who finally confirmed this relationship (although some would debate whether this was confirmed or not) by comparison of 18S rRNA sequences. Additional supporting spermatological evidence has accumulated since that publication (e.g., Storch, 1984; Storch and Jamieson, 1992). Storch and Jamieson (1992) concluded that "a sister-group relationship of pentastomids and Branchiura . . . is confirmed" and that "the sperm of the pentastome-branchiuran assemblage appear to be the most highly evolved of the flagellate crustacean sperm." Some modern invertebrate texts now treat the pentastomids as crustaceans (e.g., Brusca and Brusca, 1990; Ruppert and Barnes, 1994). Brusca and Brusca (1990) mention additional evidence such as similarities in the type of embryogenesis, cuticular fine structure, and arrangement of the nervous system.

Nevertheless, the amazing discovery of fossils from Middle Cambrian limestones that are extremely similar to extant pentastomes (Walossek and Müller, 1994; Walossek et al., 1994) would seem to cast doubt on placing them within the Crustacea (see discussions in Walossek and Müller, 1994, 1998; also Almeida and Christoffersen, 1999) and certainly would argue against their being maxillopods. If these fossils are indeed related to modern-day pentastomids (an issue we feel is not yet settled, but see Almeida and Christoffersen, 1999, for a dissenting opinion), then this finding would dispel any notion that the pentastomes are a recently derived group. Walossek and Müller (1994) make the point that, if pentastomids are related to branchiurans, then the morphology of the two groups as well as their modes of development have differed markedly for more than 500 million years, such that present day similarities of their sperm morphology might seem to carry less weight. If the Cambrian fossils are indeed pentastomids—appearing hundreds of millions of years before most of their present day hosts were on the scene—we must rethink whether we can accept such a major divergence in body plan so soon after the Crustacea itself appears in the fossil record. Thus, our inclusion of them here represents an acceptance of the available molecular and sperm morphology data (for additional molecular support, see Garey et al., 1996, and Eernisse, 1997) over apparently sound fossil evidence to the contrary; this may prove to be an error. Brusca (2000) suggests a way to reconcile the issues if early pentastomids were parasites of early fish-like vertebrates as represented by the conodonts, many of which were present in the Cambrian.

The classification we follow for the pentastomids is from Riley (1986; see also Riley et al., 1978). This classification has been questioned recently by Almeida and Christoffersen (1999), who do not consider pentastomes to be crustaceans. Almeida and Christoffersen suggest, based on a cladistic analysis of available genera, the recognition of the Raillietiellida as a new order to contain their new family Raillietiellidae (for the genus *Raillietiella*), the recognition of the Reighardiida as a new order to contain the family Reighardiidae, and the dissolution of the family Sambonidae. Additionally, the Porocephalida was partitioned by them into two superfamilies. We have not followed the Almeida and Christoffersen (1999:702) classification here.

Authority for the taxon name Pentastomida was somewhat difficult to decipher. Riley (pers. comm.) informs us that the name "Pentastomum" was first employed by Rudolphi (1819) to refer to a single species, and several workers (e.g., Almeida and Christoffersen, 1999) credit the taxon name Pentastomida to Rudolphi. We have been unable to locate a work by Rudolphi in 1819 and suspect that Rudolphi, 1809, was the intended reference, as Rudolphi described the genus *Pentastoma* and used the group name Pentastomata in this 1809 work (L. Holthuis, pers. comm.). Diesing (1836) first used it (as Pentastoma) for the entire group, although the rank was not given. Elevation to phylum status was not suggested until 1969 (Self, 1969), although his evidence and reasoning were flawed (Riley, pers. comm.). Prior to that, there were various spellings and ranks assigned (e.g., by Heymons, 1935; Fain, 1961; and others; see Riley, 1986). Thus, because Diesing was the first to use the name Pentastoma for the entire assemblage, we have attributed the authorship of the Pentastomida to him.

Riley (1986) also was of the opinion that pentastomids were allied with arthropods and probably with crustaceans, noting that "the available evidence overwhelmingly indicates that pentastomids are euarthropods and, more specifically, that their affinities are closer to crustaceans than uniramians." More recently, however, he has indicated that the return to the status of separate phylum is probably warranted (pers. comm., 1998). Riley's (1986) classification (his table 1), which we have followed, recognized nine families in two orders. Two suborders of the Porocephalida are mentioned in Riley's text, but he chose not to recognize them in his table, and we have followed his lead.

The inclusion of pentastomids among the Crustacea takes the known morphological diversity and lifestyle extremes of the Crustacea—already far greater than for any other taxon on earth—to new heights. How many other predominantly marine invertebrate taxa can claim to have representatives living in the respiratory passages of crocodilians, reindeer, and lions?

## SUBCLASS MYSTACOCARIDA, ORDER MYSTACOCARIDIDA

To our knowledge, there have been no suggested changes in the classification of, or in our understanding of the phylogeny of, the mystacocarids since Bowman and Abele (1982). The subclass continues to be represented by a single extant order (Mystacocaridida) and family (Derocheilocarididae). In their review of crustacean relationships based on 18S rDNA, Spears and Abele (1997) noted, within the maxillopodan groups, that "the long branch leading to the first lineage, the Mystacocarida, indicates extensive divergence relative to other crustaceans." Schram et al. (1997) suggest a mystacocarid + copepod lineage; a relationship with copepods has also been suggested by Boxshall and Huys (1989) and Walossek and Muller (1998). The group was most recently reviewed by Boxshall and Defaye (1996) and Olesen (2001).

## SUBCLASS COPEPODA

What could have been a truly daunting task for us has been made considerably easier by the relatively recent publication of *Copepod Evolution* by Huys and Boxshall (1991), by Damkaer's (1996) list of families of copepods (along with their type genus), and by three recent treatments of copepods by Razouls (1996, free-living copepods), Raibaut (1996, parasitic copepods), and Razouls and Raibaut (1996, phylogeny and classification). Our acceptance of the Huys and Boxshall classification resulted in 26 families that have been added, while 18 families recognized by Bowman and Abele have been replaced, resulting in a net gain of 8 families. Additional families have been described or recognized since then (listed below). Huys and Boxshall (1991) proposed some rather sweeping changes in some of the higher taxonomic levels as well. Indeed, most of the suborders and superfamilies appearing in the Bowman and Abele (1982) list have been suppressed. This tack was taken also by Damkaer (1996), although he does not cite Huys and Boxshall. Where the two classifications differ, we tended to follow Huys and Boxshall (1991), and readers are referred to that tome for arguments underlying these changes. However, we must also point out that not everyone has accepted the changes suggested by Huys and Boxshall (1991) (see especially the critique by Ho, 1994a). Indeed, Huys continues to use the superfamily concept in some instances (see Huys and Lee, 1999, for the Laophontoidea) even though it was not used in Huys and Boxshall (1991). W. Vervoort (pers. comm.) reminds us that "a subdivision of a subclass the size as that of the Copepoda will always remain a matter of personal choice," and indeed some of the changes advocated by Huys and Boxshall have been corrected by these same authors in subsequent personal communications, as noted below. These changes represent not a capricious nature but our constantly changing understanding of a tremendously diverse group of organisms.

Just prior to the publication of Huys and Boxshall's book, Ho (1990) presented a cladistic analysis of the orders of the copepods. The results of that analysis differ in several significant ways from the classification of Huys and Boxshall (and thus from our classification). For example, Ho (1990) recognized a gymnoplean clade that included the Platycopioidea and Calanoidea, and this clade was the sister group to the remaining copepod orders. In contrast, Huys and Boxshall (1991) treated the Platycopioidea as being outside of the Gymnoplea. There are other differences as well, such as the placement of the monstrilloids and cyclopoids. Ho (1990) consistently placed these taxa near each other, whereas Huys and Boxshall (1991) separate them in their classification, at least implying that they are not closely related. In his subsequent critique of the Huys and Boxshall (1991) phylogeny, Ho (1994a) pointed out an alternative phylogeny where the Misophrioida was depicted as the sister group to the remaining seven orders of the Podoplea. For an in-depth review of recent attempts at producing copepod phylogenies, interested readers should consult Huys and Boxshall (1991) and the critique by Ho (1994a). A more recent molecular study (Braga et al., 1999) of relationships among the Poecilostomatoida, Calanoida, and Harpacticoida yielded somewhat different results, with the Poecilostomatoida depicted as basal to the calanoids and harpacticoids, in contrast with the abovementioned morphology-based hypotheses.

The review of copepod phylogeny and classification presented by Razouls and Raibaut (1996) (based in part on Boxshall, 1983, 1986; Boxshall et al., 1984; Por, 1984) recognizes 10 orders of copepods, as did Huys and Boxshall (1991). However, Razouls and Raibaut (1996) did not list the orders under superorders or subclasses, preferring instead to treat each order separately and refrain from phylogenetic hypotheses (although they reproduce the "tree" of important events in the evolution of copepods from Boxshall, 1986). Also, the list of accepted families within each order is not always the same in the two treatments. The included families are not always given by Razouls and Raibaut (1996), and there are differences in the names and dates assigned to some of the families. It is also apparent that some phylogenetic information may be forthcoming from detailed studies of copepod developmental (naupliar and copepodid) stages (e.g., see Dahms, 1990, 1993), but the data to date are preliminary and incomplete (Dahms, 1990).

M. Grygier informs us (pers. comm.) that the correct date for the many copepod taxa named by Giesbrecht should perhaps be 1893 rather than 1892; he refers to Scott's (1909) note in the Siboga Expedition (a note added to the entry for Giesbrecht's Naples volume in the reference list of Scott, 1909). We have not seen Scott's 1909 reference list, but the date 1893 has been confirmed by W. Vervoort (pers. comm.), who additionally notes that Scott was a contemporary of Giesbrecht and that there is therefore "no reason at all to doubt [his] accuracy." We have thus used this date (1893) instead of the often-used 1892.

Publications describing or recognizing additional families subsequent to Bowman and Abele (1982), some of which appeared too late for inclusion in (or subsequent to) Huys and Boxshall (1991), are listed in the following sections on copepod orders.

## ORDER PLATYCOPIOIDA

This newly recognized order (established by Fosshagen, 1985, in Fosshagen and Iliffe, 1985) is based on the family Platycopiidae Sars, 1911, and currently contains only that family and its four genera (*Platycopia, Nanocopia, Sarsicopia,* and *Antrisocopia*). Because Sars established the family Platycopiidae, an argument could be made that Sars should be the name associated with the higher taxon as well, although most workers credit Fosshagen (correctly) and/or Fosshagen and Iliffe (1985).

## ORDER CALANOIDA

Publications with newly described calanoid taxa include Fosshagen and Iliffe (1985; Boholinidae), Suarez-Morales and Iliffe (1996; Fosshageniidae), Ohtsuka, Roe, and Boxshall (1993; Hyperbionychidae), and Ferrari and Markhaseva (1996; Parkiidae). Suarez-Morales and Iliffe (1996) also erected a superfamily, the Fosshagenioidea, to accommodate their new family Fosshageniidae, but in keeping with our decision to follow the Huys and Boxshall (1991) classification, which avoids superfamilies, we have not included that taxon, instead listing the Fosshageniidae alphabetically among the other calanoid families. The family name Phyllopodidae has been replaced (because an older use of the name *Phyllopus* was suppressed only for purposes of synonymy and not homonymy; G. Boxshall, pers. comm.), and the family name erected to replace it is Nullosetigeridae (Soh et al., 1999). The very similar spelling of the families Pseudocyclopidae and Pseudocyclopiidae, pointed out earlier by some readers as a possible error, is in fact correct and results from the former being based on the genus *Pseudocyclops* Brady while the latter is based on the genus *Pseudocyclopia* Scott (G. Boxshall, pers. comm.). Park (1986) presented a brief discussion of calanoid phylogeny (based largely on that of Andronov, 1974); more recently, Braga et al. (1999) examined relationships among calanoid superfamilies using 28s rRNA data.

## ORDER MISOPHRIOIDA

Two new families of misophrioidans, the Palpophriidae and Speleophriidae, both comprising genera found in anchialine habitats, were described by

Boxshall and Jaume (2000, see also 1999). The palpophriids and misophriids constitute a clade that is the sister group to the Speleophriidae (Boxshall and Jaume, 1999).

## ORDER CYCLOPOIDA

Papers with new cyclopoid taxa include Boxshall (1988; Chordeumiidae), Ho and Thatcher (1989; Ozmanidae [of interest because this family is based on a new genus and species from a freshwater snail, making it, according to the authors, the "first parasitic copepod ever recorded from a freshwater invertebrate"]), da Rocha and Iliffe (1991; Speleoithonidae), and Ho et al. (1998; Fratiidae). The family Thespesiopsyllidae has been removed, as it is an objective synonym of Thaumatopsyllidae (see McKinnon, 1994). The family Mantridae, originally placed in the Poecilostomatoida, was transferred to the Cyclopoida by Huys (1990d).

We initially removed from the cyclopoids the Botrylophyllidae and Buproridae, following Huys and Boxshall (1991). Illg and Dudley (1980) recognized these as subfamilies of the Ascidicolidae (along with five other subfamilies), and Huys and Boxshall (1991) followed that arrangement. However, Huys (pers. comm.) has suggested that the Buproridae (and also the Botrylophyllidae; see below) should be reinstated. G. Boxshall (pers. comm.) also feels that the Ascidicolidae, as constituted, "is too heterogeneous and the Buproridae at least should be accorded separate family status." However, the situation with the Botrylophyllidae is more problematic, one problem being that it is a junior synonym of the Schizoproctidae (Illg and Dudley, 1980; G. Boxshall, pers. comm.); Boxshall (pers. comm.) feels that most, but not all, of the seven ascidicolid subfamilies recognized by Illg and Dudley (1980) "will eventually be given full family status." Thus, we have reinstated the Buproridae but not the Botrylophyllidae. The former families Enterocolidae, Enteropsidae, and Schizoproctidae were also reduced to subfamilies of the Ascidicolidae by Illg and Dudley (1980), according to J.-S. Ho (pers. comm.). The family Cucumaricolidae was transferred here from the Poecilostomatoidea following Huys and Boxshall (1991), among other such changes (see their book). Other changes to the Bowman and Abele (1982) list include the removal of the Doropygidae (long known to be a synonym of the Notodelphyidae) and the Namakosiramiidae (a synonym of the harpacticoid family Laophontidae) (J.-S. Ho, pers. comm.; G. Boxshall, pers. comm.). Ho (1994b) discussed cyclopoid phylogeny (based on cladistic analysis of the 10 families known at that time) and concluded that parasitism had arisen twice in the group.

## ORDERS GELYELLOIDA AND MORMONILLOIDA

The order Gelyelloida was established by Huys (1988) for the family Gelyellidae, treated in the past as a harpacticoid family and listed as "infraorder incertae cedis" by Bowman and Abele (1982:11). The Mormonilloida is unchanged, consisting still of the single family Mormonillidae.

## ORDER HARPACTICOIDA

Papers describing new harpacticoid taxa (or elevating former subfamilies) include Huys (1990a, Adenopleurellidae; 1990b, Hamondiidae, Ambunguipedidae; 1990c, Cristacoxidae, Orthopsyllidae), Por (1986, Argestidae, Huntemanniidae, Paranannopidae [revised by Huys and Gee, 1996], Rhizothricidae [splitting the polyphyletic Cletodidae]), Fiers (1990, Cancrincolidae), Huys and Willems (1989, Laophontopsidae, Normanellidae; see also Huys and Lee, 1999), Huys and Iliffe (1998, Novocriniidae), Huys (1988, Rotundiclipeidae), Huys (1993, Styracothoracidae), and Huys (1997, Superornatiremidae). Huys and Lee (1999) elevated to family level the Cletopsyllinae, formerly a subfamily of the Normanellidae (following Huys and Willems, 1989). The Paranannopidae established by Por (1986) was relegated to a subfamily of the Pseudotachidiidae by Willen (1999); the Pseudotachidiidae was formerly a subfamily of the Thalestriidae. Huys et al. (1996) referred to this assemblage (the Paranannopidae) as the Danielsseniidae Huys and Gee because Paranannopidae was based on an unavailable genus name. Thus, the family Paranannopidae (= the Danielsseniidae of Huys et al., 1996) does not appear in our list, as it is considered a subfamily of the Pseudotachiidae following Willen's (1999) preliminary study. The subfamily Leptastacinae Lang was upgraded to a family by Huys (1992). The family Gelyellidae, treated by Bowman and Abele (1982) as a harpacticoid family, was transferred to its own order, Gelyelloida, by Huys (1988). Relationships among the laophontoidean families were addressed by Huys (1990b) and Huys and Lee (1999).

Arbizu and Moura (1994) found the family Cylindropsyllidae polyphyletic and elevated the former subfamily Leptopontiinae to family level (family Leptopontiidae). Although they also suggested that the family Cylindropsyllidae should be relegated to a subfamily of the Canthocamptidae, we have retained the family Cylindropsyllidae for now (and on the advice of R. Huys, pers. comm.).

## ORDER POECILOSTOMATOIDA

Papers describing new poecilostomatoid taxa include Humes (1986, Anthessiidae), Humes and Boxshall (1996, Anchimolgidae, Kelleriidae, Macrochironidae, Octopicolidae, Synapticolidae, Thamnomolgidae), Avdeev and Sirenko (1991, Chitonophilidae [incomplete description; tentative placement in the Poecilostomatoida is based on pers. comm. from W. Vervoort, A. Humes, and G. Boxshall]), Ho (1984, Entobiidae, Spiophanicolidae), Humes (1987, Erebonasteridae), Marchenkov and Boxshall (1995, Intramolgidae), Huys and

Böttger-Schnack (1997, Lubbockiidae), Lamb et al. (1996, Nucellicolidae), Boxshall and Huys (1989b, Paralubbockiidae), Ho and Kim (1997, Polyankylidae). The family Phyllodicolidae was transferred here from the cyclopoids by Huys and Boxshall (1991) (it still appears as a cyclopoid in Damkaer, 1996).

Also within the poecilostomatoids, the Lernaeosoleidae was elevated (from the Lernaeosoleinae Yamaguti, 1963) by Hogans and Benz (1990). The family Amazonicopeidae proposed by Thatcher (1986) has not been recognized; it is thought to be a synonym of the Ergasilidae by G. Boxshall (pers. comm.) and J. Ho (pers. comm., and citing Amado et al., 1995). The family Anomopsyllidae (included in Bowman and Abele, 1982, and in Huys and Boxshall, 1991) is not listed here. According to G. Boxshall (pers. comm.), "the genus *Anomopsyllus* was included in the Nereicolidae by Stock (1968), the family Anomopsyllidae thus becoming a junior synonym of the Nereicolidae." Laubier (1988) (unfortunately overlooked by Huys and Boxshall, 1991) described both sexes of the genus and confirmed that it is a nereicolid. The family Vaigamidae proposed by Thatcher and Robertson (1984) also is not included here, as it was shown to be a synonym of the Ergasilidae by Amado et al. (1995). The Nucellicolidae, although retained for now, may prove to be a junior synonym of the Chitonophilidae (R. Huys, pers. comm.). Finally, the family Micrallectidae has been established recently (Huys, 2001) to accommodate poecilostomatoid genera associated with pteropods.

Ho (1984) suggested phylogenetic relationships among the nereicoliform families, indicating three main lines of evolution. Later, Ho (1991) conducted a more thorough analysis of the 47 known poecilostomatoid families, which remains the most in-depth study of poecilostomatoid relationships while at the same time being somewhat preliminary in nature. Relationships of 10 poecilostomatoid families (in the lichomolgoid complex) are presented by Humes and Boxshall (1996). Unfortunately, we could not follow their suggestions here because of the absence of knowledge concerning the other (nonlichomolgoid) poecilostomatoid families.

### ORDER SIPHONOSTOMATOIDA

Papers describing new siphonostomatoid taxa include Izawa (1996, Archidactylinidae [questionable, as this is an incomplete description]), Humes and Stock (1991, Coralliomyzontidae), and Humes (1987, Ecbathyriontidae). The Herpyllobiidae (treated as siphonostomes by Huys and Boxshall, 1991) have been removed to the Poecilostomatoidea (R. Huys, pers. comm.). Two new families, Dichelinidae and Codobidae, have been proposed recently for siphonostomatoid genera parasitic on echinoderms (Boxshall and Ohtsuka, 2001), and the family Scottomyzontidae (erected for *Scottomyzon gibberum,* a symbiont of the asteroid *Asterias rubens*) was established by Ivanenko et al. (2001).

### ORDER MONSTRILLOIDA

With the transfer of the Thaumatopsyllidae to the Cyclopoida (Huys and Boxshall, 1991; Grygier, pers. comm.), the order Monstrilloida has been reduced to a single family, Monstrillidae, which now is credited to Dana rather than to Giesbrecht following ICZN Opinion 1869 (M. Grygier, pers. comm.).

There have also been many additional changes to the list of copepod families that are not detailed here—including additions, deletions, reinstatements of older families, corrected spellings and authors, etc.—suggested by various workers, mostly Ju-Shey Ho, Arthur Humes, H.-E. Dahms, G. Boxshall, and Rony Huys. In some cases, we did not ask for a published reference, instead taking these workers at their word (and also because in some cases the suggestion has not been published).

## CLASS OSTRACODA

This section received extensive input from Dr. Anne Cohen, and our treatment of this group is in many ways based on her impressive knowledge of this taxon. Major references included Morin and Cohen (1991), Martens (1992), Whatley et al. (1993), Hartmann and Guillaume (1996), Martens et al. (1998), and Cohen et al. (1998).

Many previous workers have considered ostracodes to be a subclass of the Maxillopoda. The strongest reason for including ostracodes among maxillopods is, apparently, the presence in ostracodes of a naupliar eye with three cups and tapetal cells between the sensory and pigment cells (e.g., see Elofsson, 1992; Huvard, 1990; and earlier papers cited in these works). This feature is found also in the Thecostraca, Branchiura, and Copepoda, and for this reason, Schram (1986), Brusca and Brusca (1990), and others have placed ostracodes within the Maxillopoda (see also discussions in Grygier, 1983a; Boxshall, 1992; Elofsson, 1992; Cohen et al., 1998). Schram (pers. comm., and citing K. Schultz, *Das Chitinskelett der Podocopida und der Frage der Metamerie dieser Gruppe,* doctoral dissertation, University of Hamburg, which we have not seen) informs us that an additional apomorphy that argues for inclusion of ostracodes within the Maxillopoda is the location of the gonopods. Swanson's (1989a, b, 1990, 1991) discovery of living specimens of the primitive ostracode genus *Manawa* (family Punciidae) caused him to suggest the inclusion of ostracodes within the Maxillopoda as well. Cohen et al. (1998) note the following "perhaps homologous morphological characters": a medial naupliar eye that has three cups and a tapetal layer (present in most Myodocopida and in many Podocopida), and overall reduction in body size and limb number.

However, other workers are quick to point out

that reduction in body segmentation has occurred independently as a functional adaptation in many different and unrelated crustacean taxa and that the unique features of the Ostracoda argue for their recognition as a separate class (see especially discussions in Newman, 1992; Boxshall, 1992; Wilson, 1992). Treatment of ostracodes as a subclass of the Maxillopoda has additional problems as well. Wilson (1992) could not find support for placing the former within the latter based on morphological grounds (although Schram and Hof, 1998, point out errors in Wilson's analysis that, if corrected, would indeed group ostracodes with one cluster of Maxillopoda). Abele et al. (1992) rejected the inclusion of ostracodes in the Maxillopoda on molecular grounds. Spears and Abele (1997) suggest the possibility that, based on molecular data, both Ostracoda and Maxillopoda might be paraphyletic.

There is also some evidence, both morphological and molecular, that the two major groupings of the Ostracoda (Myodocopa and Podocopa) may not constitute a monophyletic assemblage (e.g., see Vannier and Abe, 1995; Spears and Abele, 1997). On the other hand, Cohen et al. (1998), based on the many similarities between these two groups, "regard it more parsimonious and useful to assume that they do." This older view—that ostracodes are monophyletic—has been adopted here and is in fact held by a majority of current workers in the field. The assignment of ostracodes to a group "Entomostraca" (which included, in addition to ostracodes, the Branchiopoda, Cirripedia, Branchiura, and Phyllocarida) by McKenzie et al. (1983) was clearly an unsupported departure (see also discussions on Branchiopoda and Phyllocarida and notes on Entomostraca under the general heading Crustacea).

A modified version of the classification of the Ostracoda used by Whatley et al. (1993), which will be the basis for the classification used in the upcoming revision of the Treatise on Invertebrate Paleontology ("more or less," according to Whatley, pers. comm.; R. Kaesler, pers. comm.), was sent to us by R. Whatley. This classification, which differs considerably from what was proposed by McKenzie et al. (1983) and also from the classification used by Hartmann and Guillaume (1996), has been followed fairly closely. Differences include the spelling of the endings of superfamilies. We use the ICZN-recommended ending "–oidea" (which in the latest (fourth) edition of the International Code of Zoological Nomenclature is mandatory rather than a recommendation; ICZN, 1999, article 29.2). Whatley, in one of the more interesting responses we received, has indicated that the "-oidea" spelling is an "attempted imposition" by the ICZN. Kaesler (pers. comm.) and Whatley (pers. comm.) note that ostracodologists prefer to think of the higher groups as superfamilies rather than as suborders and are also more accustomed to the use of the ending "-acea" for superfamilies and thus are more familiar with, and prefer, the concept of a superfamily Bairdiacea as opposed to a superfamily Bairdioidea or suborder Bairdiocopina. On the spelling of superfamily names, however, the ICZN recommendation (ICZN, 1999, fourth edition, article 29.2) is rather clear: "The suffix -OIDEA is used for a superfamily name, -IDAE for a family name, -INAE for a subfamily name . . ." etc. And it appears to us that it is primarily the paleontologists (who are, we admit, the majority of the ostracodologists) rather than neontologists who prefer (and use) the "-acea" ending for superfamilies (e.g., see Martens, 1992, and Martens et al., 1998, for living freshwater ostracode superfamilies, all of which are spelled according to ICZN recommendation 29.A [now 29.2]). As Martens et al. (1998: 41) explain in a note to accompany their classification, ". . . as ostracods are animals, we will follow the ICZN throughout this book."

Thus, we have followed the ICZN recommendation (as did Bowman and Abele, 1982, and Schram, 1986) for spellings of superfamilies (e.g., Bairdioidea, not Bairdiacea). Whatley (pers. comm.) also feels that, relative to the Podocopida, the Myodocopa is probably "one hierarchical level too high." Whatley (pers. comm.) considers his own arrangement (Whatley et al., 1993) "old fashioned but acceptable to people who actually work on the group," a justification that we feel is baseless but that, at the moment, faces nothing in the way of a serious alternative classification. Martens (1992) and Martens et al. (1998) appear to base their decisions more on shared derived characters and more often than not employ characters of the entire animal (as opposed to those of the shell only). Consequently, we have followed their lead for the names, spellings, and arrangement of the superfamilies and families of the freshwater families as far as was possible (not all families are treated in those works). Thus, although Whatley would remove the superfamilies Macrocypridoidea and Pontocypridoidea (placing their families among the Cypridoidea), we have maintained these groupings following Martens (1992) and Martens et al. (1998). Whatley (pers. comm.) also feels that the family Saipanettidae (= Sigilliidae; see later) is no more than a subfamily of the Bairdiidae, whereas Martens (1992) recognized a separate superfamily, the Sigillioidea Mandelstam, to accommodate this unusual group, and here again we have followed Martens (1992).

Whatley (pers. comm.) and Whatley et al. (1993) also place the unusual and primitive family Punciidae in the Platycopida (he considers *Manawa* to be a member of the Cytherellidae), indicating that there are still no living members of the Palaeocopidae. Martens et al. (1998) also feel that there are no living palaeocopids, which also supports transfer of the punciids. We have followed Whatley's advice in moving the punciids to the Platycopida (although they appear to share no unique characters with platycopids and differ in many respects), but we have retained them in their own family, the Punciidae, as we are not aware of any publications that demonstrate that they belong among the cytherellids. Possibly a better solution would have been to list them as *incertae sedis* for now.

Although there have been rearrangements of the Ostracoda, there have been surprisingly few higher taxa described or recognized since Bowman and Abele (1982) and Cohen (1982). The fossil bradoriids and the "phosphatocopines" of Sweden's Upper Cambrian "Orsten" fauna are no longer considered true ostracodes. Walossek and Müller (1998, in Edgecombe) hypothesize that, although phosphatocopines are not crown group crustaceans (their "Eucrustacea"), they may be the sister taxon to this group.

### SUBCLASS MYODOCOPA

Arrangement of families in the Myodocopa follows Kornicker (1986:178), which in turn was based largely on McKenzie et al. (1983), although some of the higher taxon spellings have been changed for consistency. The suborder Cladocopina may be deserving of status as a separate order (A. Cohen, pers. comm.), although this step has not been taken here (see also Kornicker and Sohn, 1976, who first suggested the inclusion of the Cladocopina and Halocypridina within the Halocyprida).

### SUBCLASS PODOCOPA

The superfamilies Bairdioidea and Cytheroidea have been elevated to suborders, with spelling changed to Bairdiocopina and Cytherocopina (respectively) (following Martens, 1992, and A. Cohen, pers. comm.). Alexander Liebau (pers. comm.) informs us that the Cytherocopina has been divided by him (Liebau, 1991; not seen by us) into two infraorders, the Nomocytherinina (which includes species showing epidermal cell constancy reflected by mesh constancy of reticulate sculptures) and the Archaeocytherinina, containing the paraphyletic remaining cytherocopines. We have not used this division here. Within the Cytheroidea, we have used the list of families supplied by R. Whatley (pers. comm.), based in part on Whatley et al. (1993) and on his anticipation of the Ostracoda section of the next edition of the Treatise on Invertebrate Paleontology (Whatley, pers. comm.; R. Kaesler, pers. comm.). The family Bonaducecytheridae McKenzie has been removed (R. Maddocks, pers. comm.). The superfamily Terrestricytherioidea and its sole family, the Terrestricytheridae, have been removed; Martens et al. (1998), citing Danielopol and Betsch (1980), note that *Terrestricypris* is a modified member of the Candonidae (the spelling of which has been corrected from Candoniidae; R. Maddocks, pers. comm.).

The suborder Metacopina now contains only fossils and thus has been removed from our classification, as the Darwinulocopina has now been established by Sohn (1988) to accommodate the family Darwinulidae (A. Cohen, pers. comm.). The former superfamily Cypridoidea is now treated as a suborder, Cypridocopina Jones (Martens et al., 1998). The family Paracyprididae has been removed; this group also is now thought to be a subfamily of the Candonidae (Martens et al., 1998). The Cypridopsidae has been removed (Martens et al., 1998). The family Saipanettidae, formerly in the superfamily Healdioidea (which has been removed), also has been removed. The Saipanettidae was found to be a junior synonym of the Sigilliidae, an extant family reviewed recently by Tabuki and Hanai (1999). Spelling of the Sigilliidae was initially given as Sigillidae by Tabuki and Hanai (1999); we have corrected it based on the spelling of the genus *Sigillium*. The Sigilliidae is now treated as a member of the superfamily Sigillioidea (see Tabuki and Hanai, 1999; spelling emended from Sigilloidea; R. Maddocks, pers. comm.), which in turn has been placed in its own suborder, the Sigilliocopina (see Martens, 1992). Martens (1992) originally suggested recognition at the infraorder level, as "infraorder 3, 'Sigillioidea.' " The spelling we use for the suborder was first employed by Cohen et al. (1998).

## CLASS MALACOSTRACA

Because of their size and numbers, malacostracans have been the subject of a huge number of classificatory and phylogenetic studies employing morphological characters, molecular characters, or both. For the most part, there seems to be agreement that the Malacostraca itself is a monophyletic grouping (e.g., see Hessler, 1983; Dahl, 1983a, b, 1991; Mayrat and Saint Laurent, 1996; Shultz and Regier, 2000; Watling et al., 2000; Richter and Scholtz, in press), although differing opinions can certainly be found. There is considerably less agreement concerning the constituencies and relationships of the various groupings of the Malacostraca, and these topics are the subject of a vast body of literature (much of which was reviewed recently by Richter and Scholtz, in press). Attempts to place phyllocarids outside the Malacostraca have largely been shown to be misguided (see below). We have tried to refer readers to the salient papers that offer arrangements that differ from our own in the individual sections that follow.

### SUBCLASS PHYLLOCARIDA, ORDER LEPTOSTRACA

The status of the subclass Phyllocarida (which includes only one extant order, the Leptostraca) as true malacostracans is now fairly well accepted. Arguments can be found in Dahl (1987), in rebuttal to Schram (1986), who had been in favor of resurrecting the older term Phyllopoda to include branchiopods, cephalocarids, and leptostracans (see also Rolfe, 1981; Dahl, 1992; Martin and Christiansen, 1995a; Spears and Abele, 1999; Richter and Scholtz, in press; but see also Ferrari, 1988, for a rebuttal of Dahl's criticism). Inclusion of leptostracans within the Malacostraca has been further supported by molecular evidence (rDNA data summarized in Spears and Abele, 1997, 1999; see also Shultz and Regier, 2000, for EF-1α and Pol II data). Hessler (1984) established the family Nebaliopsidae in recognition of the great differences setting the genus *Nebaliopsis* apart from other leptostracans, thereby doubling the number of recognized families of the extant phyllocarids. However, J. Olesen (1999b, and pers. comm.) finds that, depending upon the choice of outgroups (and characters) used in cladistic analyses of the group (based on descriptions in the literature), there is still some room for doubt as to whether Nebaliidae is monophyletic or paraphyletic (with *Nebaliopsis* nested within the other nebaliacean genera). Most recently, Walker-Smith and Poore (2001) have erected a third family, Paranebaliidae, to contain the genera *Paranebalia* and *Levinebalia* (the latter of which was described by Walker-Smith, 2000).

Our treatment of the Phyllocarida follows Hessler (1984), Martin et al. (1996), Dahl and Wägele (1996), and our PEET web page for Leptostraca (URL http://www.nhm.org/~peet/) in recognizing two extant families (see Rolfe, 1981, for extinct phyllocarids) plus the recently established family Paranebaliidae following Walker-Smith and Poore (2001). Most authors in the past have credited the family Nebaliidae to Baird (1850). However, according to L. Holthuis (pers. comm.), Samouelle (1819:100) mentioned "Fam. VI. Nebaliadae" [sic] in his "Entomologist's Useful Compendium," which of course predates Baird's (1850) work. Thus, we have attributed the family Nebaliidae to Samouelle, 1819.

## SUBCLASS HOPLOCARIDA, ORDER STOMATOPODA

Several workers, today and in the past (examples include Hessler, 1983; Scholtz, 1995; Richter and Scholtz, in press), have considered the hoplocarids to be members of the Eumalacostraca, a placement that has been used often and in some textbooks as well (e.g., Brusca and Brusca, 1990). However, we have retained their placement as a separate subclass within the Malacostraca pending further exploration of this question (see review by Watling et al., 2000). Our treatment of the hoplocarids as separate from the other Eumalacostraca also is consistent with some (admittedly weak) molecular evidence (see Spears and Abele, 1997, 1999b) and with cladistic analyses based mostly on fossil taxa (e.g., Hof, 1998a, b; Hof and Schram, 1999). Schram (1971, 1986) had argued earlier for separate status of the hoplocarids as well. Spears and Abele (1997) could state only that the position of the "Hoplocarida relative to the Eumalacostraca is equivocal" (low bootstrap value) based on rDNA sequence data, and thus they were "unable to determine whether hoplocarids represent a separate, independent malacostracan lineage with taxonomic rank (subclass) equivalent to that of phyllocarids and eumalacostracans." Their subsequent paper (Spears and Abele, 1999b) seems (to us) to indicate somewhat stronger evidence that hoplocarids are not eumalacostracans, but the authors are suitably cautious in not saying so. Without firm indications that we should do otherwise, we have maintained separate status for the Hoplocarida and Eumalacostraca. Although a thorough cladistic analysis of fossil and extant crustacean taxa by Schram and Hof (1998) resulted in a tree that showed hoplocarids arising from somewhere within the Eumalacostraca, these authors also noted that forcing the hoplocarids into a "sister group" position to the Eumalacostraca increased tree length by only 1%. Other workers (e.g., Watling, 1999a), recognizing how very derived the stomatopods are, place them in the Eumalacostraca as the sister taxon to the Eucarida. Most recently, Richter and Scholtz (in press) suggested that hoplocarids occupy a basal position within the Eumalacostraca. Thus, placement of the hoplocarids continues to be an unresolved issue, but we felt that the weight of the evidence placed them outside, rather than within, the Eumalacostraca. Scholtz (pers. comm.) additionally suggests that our crediting the name Eumalacostraca to Grobben is therefore incorrect, as Grobben included the hoplocarids among the Eumalacostraca (but see earlier notes on names, dates, and the ICZN).

Within the Hoplocarida, most of our changes are based on the catalog provided by H.-G. Müller (1994) and on Manning (1995), and our final arrangement of families and superfamilies follows the recent cladistic analysis by Ahyong and Harling (2000). Publications that describe or recognize families or higher taxa of stomatopods subsequent to Bowman and Abele (1982) include Manning (1995, Indosquillidae, Parasquillidae, Heterosquillidae), Manning and Bruce (1984, Erythrosquillidae [for which the superfamily Erythrosquilloidea was later created by Manning and Camp, 1993]), Manning and Camp (1993, Tetrasquillidae), Moosa (1991, Alainosquillidae), and Ahyong and Harling (2000, superfamilies Eurysquilloidea and Parasquilloidea).

Concerning phylogeny within the Hoplocarida, there is recent evidence from several laboratories that the superfamily Gonodactyloidea as presented in Bowman and Abele (1982) is not a monophyletic grouping (Hof, 1998b; Ahyong, 1997; Barber and Erdmann, 2000; Ahyong and Harling, 2000; Cappola and Manning, 1998; Cappola, 1999) and that within the gonodactyloids the eurysquillids may be paraphyletic. These same authors disagree over whether the Bathysquilloidea are monophyletic (Cappola and Manning, 1998) or not (Ahyong, 1997). A comparative study of eye design in stomatopods (Harling, 2000) also supports a nonmonophyletic Gonodactyloidea and questions the five-superfamily scheme of Müller (1994). The nonmonophyly of the Gonodactyloidea necessitates the creation of additional families and superfamilies to accommodate some of the former gonodactyloid taxa (Ahyong, 1997; Ahyong and Harling, 2000; Cappola and Manning, 1998). Cappola and Manning (1998) also suggested that a new superfamily and family (Eurysquilloidoidea, Eurysquilloididae) should be established to accommodate the former eurysquillid genus *Eurysquilloides*. We have followed the classification suggested by Ahyong and Harling (2000). According to their scheme, the families Eurysquillidae and Parasquillidae, formerly treated as members of the Gonodactyloidea, are each deserving of superfamily status, and thus they established the superfamilies Eurysquilloidea and Parasquilloidea to accommodate them. The Gonodactyloidea has been reconfigured and now contains the Alainosquillidae, Hemisquillidae, Gonodactylidae, Odontodactylidae, Protosquillidae, Pseudosquillidae, and Takuidae. The family Heterosquillidae established by Manning (1995) has been removed, as it was suggested to be a synonym of Tetrasquillidae (see Ahyong and Harling, 2000). In the most recent treatment, Ahyong (2001) synonymized the Harpiosquillidae Manning with the Squillidae; thus the Harpiosquillidae is not in our list.

Hof (1998b) recognized two main clades of extant stomatopods. One clade included most of the gonodactyloid families but excluded the alainosquillids and the eurysquillids. The second clade contained the remaining extant families and indicated possible affinities between the squilloids and lysiosquilloids and also between the bathysquilloids and erythrosquilloids. Hof (1998b) points out that, although his results are preliminary, the fact that fossils should be included when at all possible in any cladistic analysis is clear and obvious from his work. A cladistic analysis of the hoplocarids that incorporated Paleozoic taxa was presented by Jenner et al. (1998), but it did not resolve relationships within the sole extant order (their Unipeltata). In the most recent treatment, Ahyong and Harling (2000) have also suggested that the recent stomatopods have evolved "in two broad directions from the outset," corresponding roughly to the smashing and spearing types.

## SUBCLASS EUMALACOSTRACA

The concept of the Eumalacostraca as a monophyletic assemblage has not been seriously challenged, with the exception of the question of whether hoplocarids belong (see above discussion under Hoplocarida for arguments as to their inclusion or exclusion). Our classification is roughly similar to that of Bowman and Abele (1982) in recognizing the Eumalacostraca and its constituent groups, although there have been several significant rearrangements within and among those groups, as noted below (see also Richter and Scholtz, in press). Schram (1984a) reviewed characters that defined the various eumalacostracan groups recognized at that time and presented alternatives to more traditional classifications.

## SUPERORDER SYNCARIDA, ORDERS BATHYNELLACEA AND ANASPIDACEA

Monophyly of the Syncarida appears to be fairly well accepted (e.g., Schram, 1984b; Richter and Scholtz, in press). Within the Bathynellacea, we have removed the family Leptobathynellidae, as this was synonymized with the Parabathynellidae by Schminke (1973:56). Schram (1984b) credits both names (Bathynellidae, Bathynellacea) to Chappuis (1915), whereas Lopretto and Morrone (1998) credit the Bathynellidae to Grobben (as did Bowman and Abele, 1982) and the Bathynellacea to Chappuis. We have not been able to locate a paper by Grobben describing bathynellids and so have followed Schram's (1984b, 1986) lead, crediting both taxa to Chappuis (1915). The Anaspidacea remains unchanged, with four extant families.

Thus, our classification of the Syncarida and its two orders (Anaspidacea and Bathynellacea) is the same as that presented by Lopretto and Morrone (1998), where all known syncarid genera are also listed, and is essentially the same as the classification suggested earlier by Schram (1984a:196) based on a phylogenetic analysis of fossil syncarids (excluding the entirely fossil order Paleocaridacea).

## SUPERORDER PERACARIDA

We continue to recognize the Peracarida, treating it as a superorder that contains nine orders. This is mostly in keeping with Bowman and Abele (1982) and most major treatments since that time (see especially Hessler and Watling, 1999; Richter and Scholtz, in press). However, there have been suggestions made to abandon the Peracarida or at least significantly revise it (e.g., Dahl, 1983a), and the relationships among the various peracarid groups (and of peracarids to other groups of crustaceans) are very controversial. Schram (1986) advocated eliminating the Peracarida of earlier workers, feeling that it united groups that were only superficially similar. Other workers (e.g., Pires, 1987; Brusca and Brusca, 1990; Wagner, 1994; Hessler and Watling, 1999; Richter and Scholtz, in press) recognize the group, but the treatments occasionally differ as to which orders are included. Hessler and Watling (1999) review major attempts to phyletically order the peracarids, including Schram (1986), Watling (1983), Wills (1997), and Wheeler (1997), all of which have appeared subsequent to the Bowman and Abele (1982) classification. There is little agreement among these various schemes. Mysidaceans in particular are sometimes treated as one order, sometimes as the separate orders Lophogastrida and Mysida within the Peracarida, and sometimes suggested to fall outside of the Peracarida altogether. As examples, Watling (1998, 1999b) argues that mysids should fall outside the Peracarida and that the Amphipoda are deserving of status separate from all other peracarids and should constitute their own superorder as a sister group to the remaining taxa, which would then constitute a reduced Peracarida *sensu stricto*. (Interestingly, if the Mysidacea and Thermosbaenacea are removed from Watling's (1981) fig. 1, then the Amphipoda would indeed appear as the sister group to all other "true" peracaridans in that diagram.) But this is not in agreement with Wagner (1994), who depicted amphipods and isopods as closely related and depicted amphipods, isopods, cumaceans, and tanaidaceans as a monophyletic clade. Wagner (1994) also suggested affinities between the Thermosbaenacea and Mictacea and between those two groups and the Spelaeogriphacea, whereas Pires (1987) treated amphipods and mysidaceans as related taxa that were in turn the sister group to all other peracarids. In Wagner's phylogenies, the mysids (both Mysida and Lophogastrida) are shown as the sister group to the other Peracarida. Depending on where the line is drawn, Wagner's phylogeny could be used as an argument for inclusion or exclusion of the mysids within the Peracarida.

Spears and Abele (1997, 1998) have suggested, on the basis of molecular data, that the two groups of mysidaceans are not monophyletic (suggested earlier by Dahl, 1983a, and others based on morphological features), with the Lophogastrida grouping with other peracarids but with the Mysida falling outside that clade (see below). Jarman et al. (2000) also concluded (on the basis of 28S rDNA sequence data) that the Mysida and Lophogastrida are not closely related but posited the Mysida closer to the Euphausiacea. Thermosbaenaceans, treated as true peracarids by us (see arguments below and also Richter and Scholtz, in press), have in the past been treated by some workers (e.g., Bowman and Abele, 1982; Pires, 1987) as the separate order Pancarida, which we have abandoned. A more radical departure is suggested by Mayrat and Saint Laurent (1996), who suggested a phylogeny (their fig. 342) of the Malacostraca in which the peracarids are polyphyletic, with amphipods depicted as the sister taxon to all other malacostracans (except the leptostracans) and with cumaceans and mysids associated with the higher eumalacostracans. This, to us, seems unlikely. Richter (1999), after a thorough analysis of characters of the compound eyes of malacostracans, felt that "Lophogastrida and Mysida are clearly members of the Peracarida." These are only a few of the suggestions to be found in the rather confusing literature on the diverse peracarid crustaceans. The most recent coverage is a wonderful in-depth treatment of the entire Peracarida in Tome VII, fascicule IIIA of the *Traité de Zoologie* edited by J. Forest (see especially the review by Hessler and Watling, 1999).

The suggestion that the orders Spelaeogriphacea, Cumacea, Tanaidacea, and Thermosbaenacea constitute a grouping termed the "Brachycarida" that is the sister group to the Isopoda, first suggested by Schram (1981) and supported by Watling (1983, 1999b) [although note that the suggested placement of isopods and amphipods differs in these two papers], is not followed here. However, removal of the thermosbaenaceans from the "Pancarida" and grouping them with the other peracarids, which we have done, could be seen as supportive of that move (see below under order Thermosbaenacea).

Gutu (1998) and Gutu and Iliffe (1998) have suggested a novel reorganization of the peracarids, where both the spelaeogriphaceans and mictaceans would be treated as suborders of a new peracarid order, the Cosinzeneacea (Gutu, 1998). The mictacean family Hirsutiidae would be removed to the new order Bochusacea (Gutu and Iliffe, 1998). We have not followed this suggestion.

Thus, our Peracarida contains the two orders of former "mysids" treated as the separate orders Lophogastrida and Mysida (as in many earlier treatments as well; see below), plus the Thermosbaenacea, in addition to the Spelaeogriphacea, Mictacea, Amphipoda, Isopoda, Tanaidacea, and Cumacea. Additional comments on each group are given below.

## ORDER SPELAEOGRIPHACEA

To date, there are only three known extant species of this group, from South America (Brazil), South Africa, and Australia (Pires, 1987; Poore and Humphreys, 1998; see also Shen et al., 1998). Pires (1987) suggested that spelaeogriphaceans and mictaceans might be sister taxa. A recent cladistic analysis stemming from the discovery of a new genus and species from the Upper Jurassic of China (Shen et al., 1998) indicates that the Spelaeogriphacea may be paraphyletic. Although Shen et al. treat the Spelaeogriphacea as a suborder under the order Hemicaridea Schram, we have not followed that suggestion. This may change if fossil taxa are incorporated into the next edition of this classification. All species are currently considered members of a single extant family, the Spelaeogriphidae, and the group has been reviewed recently by Boxshall (1999). Gutu (1998) has suggested recently that spelaeogriphaceans and some former mictaceans (the family Mictocarididae, not the Hirsutiidae) should be treated as suborders within the newly created order Cosinzeneacea. We have not followed this suggestion, as most other workers seem to be in agreement that the two groups are deserving of separate status within the Peracarida.

## ORDER THERMOSBAENACEA

The former order Pancarida (as used in Bowman and Abele, 1982), erected to accommodate the order Thermosbaenacea, has been eliminated in light of suggestions that thermosbaenaceans are members of a redefined Peracarida clade (see discussion in Wagner, 1994; see also Monod and Cals, 1988; Cals and Monod, 1988; Spears and Abele, 1998; Richter and Scholtz, in press; and above under Peracarida). Our treatment of the Thermosbaenacea as true peracarids is in agreement with morphological interpretations (e.g., Monod, 1984; Cals and Monod, 1988; Monod and Cals, 1988, 1999) and recent molecular evidence (Spears and Abele, 1998). Other workers (e.g., Newman, 1983; Sieg, 1983a, b; Pires, 1987; A. Brandt, pers. comm.) have argued for maintaining separate status from the other peracarid groups (reviewed by Wagner, 1994). Wagner (1994), whose extensive review we followed in the current classification, also was of the opinion that there is no real justification for excluding the Thermosbaenacea from the Peracarida.

Within the Thermosbaenacea, two new families have been described since 1982: Halosbaenidae (Monod and Cals, 1988) and Tulumellidae (Wagner, 1994). The family Monodellidae was also recognized by Wagner (1994), bringing the total to four recognized extant families (up from one in Bowman and Abele, 1982). Wagner's (1994) thorough treatment also suggests some phylogenetic relationships among the thermosbaenaceans (as did Monod and Cals, 1988). The Thermosbaenidae and Monodellidae appear to be sister taxa, but the position of the Tulumellidae was undetermined, sometimes appearing as the sister group to the Halosbaenidae and sometimes as part of the thermosbaenid + monodellid clade (as in his "final proposed phylogenetic tree"; Wagner, 1994, fig. 498). Thus, we have not attempted to phyletically order the four recognized families at this time. See also the recent review by Monod and Cals (1999), where previous systematic arrangements (Cals and Monod, 1988; Wagner, 1994) are briefly discussed.

## ORDERS LOPHOGASTRIDA AND MYSIDA

Abele and Spears (1997) concluded, based on rDNA studies, that the Peracarida (including the Thermosbaenacea) is indeed a monophyletic assemblage, but only if the Mysida are excluded. Jarman et al. (2000) also would separate the Mysida, which they felt are closer to the Decapoda, from the Lophogastrida. Supporting evidence is also found in the fact that all peracarids (again including thermosbaenaceans but excluding Mysida) contain similar hypervariable regions of 18S rDNA (Spears and Abele, 1998). However, these distinctly peracarid features appear to be present in the other mysidacean group, the Lophogastrida. The inclusion of the mysids (both Mysida and Lophogastrida) in the Peracarida (e.g., as suggested most recently by Richter and Scholtz, in press) has also been questioned on morphological grounds. For example, as noted above, Watling (1998, 1999a, b) feels that the mysidaceans (i.e., both the Mysida and Lophogastrida as the taxon Mysidacea) do not belong to the Peracarida and are instead more closely allied to the eucarids. Yet both groups of the Mysidacea (Mysida and Lophogastrida) share some unique and possibly synapomorphic morphological features of the walking limbs (Hessler, 1982; see also Hessler, 1985) and foregut (De Jong-Moreau and Casanova, 2001) that suggest monophyly. Additionally, Richter (1999; see also Richter and Scholtz, in press) has shown that lophogastridans and mysidans share unique morphological components to the design of their ommatidia (although these features also are shared with Anaspidacea and Euphausiacea). The recent treatment by Nouvel et al. (1999) treats the Mysidacea as monophyletic (see also Richter, 1994, for further arguments in favor of monophyly of the Mysidacea).

Are mysidaceans paraphyletic? Is it possible that the Mysida fall outside the Peracarida *sensu stricta* but that the Lophogastrida are true peracarids (ignoring, for the moment, the larger question of whether the Peracarida itself is monophyletic)? This seems unlikely based on limb morphology (e.g., Hessler, 1982), and foregut morphology (De Jong-Moreau and Casanova, 2001), and yet other workers have noted significant differences between the Mysida and Lophogastrida on morphological (and now, it appears, on molecular) grounds. Several other workers (e.g., G. Scholtz and S. Richter, pers. comm.) commented on the distinct morphological differences between the Lophogastrida and Mysida and suggested that these taxa be elevated to ordinal status and that the former Mysidacea that contained the two be abandoned (but see also Richter, 1994, De Jong-Moreau and Casanova, 2001, and Richter and Scholtz, in press, for arguments in favor of monophyly). We have split the former order Mysidacea, elevating each of the former mysid suborders to order level, as have several other workers before us, such as Schram (1984, 1986), and Brusca and Brusca (1990:624, who note that an increasing number of specialists have begun to treat the two groups separately). This could be seen as a preliminary for removing one or both of these groups from the Peracarida, if the suggestions of Watling (1998, 1999a, b) and Spears and Abele (1998) find additional support in the future. However, we have kept the two groups within the Peracarida for now.

Taylor et al. (1998) analyzed the relationships of a group of fossil malacostracans (the Pygocephalomorpha) that are possibly allied with mysids; one of their conclusions was that the recent mysids and lophogastrids do form a clade (albeit a somewhat "confused" one). Thus, our classification is most similar to that of Brusca and Brusca (1990) in recognizing both former "mysidacean" groups as orders within the superorder Peracarida rather than as suborders within the Mysidacea (as presented by Nouvel et al., 1999). Casanova et al. (1998) examined relationships of the two lophogastrid families (Eucopiidae and Lophogastridae) based on morphological and limited molecular data. Among their conclusions was that the monogeneric eucopiids (*Eucopia*) originated from within the Lophogastridae.

Authorities and dates for some taxa in the Mysida have been changed to earlier workers and dates (e.g., Mysida Haworth and Mysidae Haworth rather than Mysida Boas or Mysida Dana) following the recommendation of L. Holthuis (pers. comm.) citing ICZN article 50(c)(i) (now 50.3.1, ICZN fourth edition, 1999). Tchindonova (1981) suggested the erection within the Mysida of the suborders Petalopthalmina and Stygiomysina as well as the tribe Amblyopsini and the family Boreomysidae (in addition to several new subfamilies, tribes, and genera; P. Chevaldonne, pers. comm.). We have not followed this suggestion.

## ORDER MICTACEA

In 1985, two groups of workers simultaneously described two new families of an entirely new order of peracarid crustaceans and then jointly described the new order (Bowman et al., 1985). The new families were the Hirsutiidae (Sanders et al., 1985) and the Mictocarididae (Bowman and Iliffe, 1985), the latter of which formed the basis of the name of the new order Mictacea. A second species of the Hirsutiidae was described from Australia by Just and Poore (1988). Although discovery of the Mictacea has prompted speculation about its phylogenetic affinities, most workers are in agreement that the group fits comfortably within the Peracarida. Thus, we include the order and its two families among the Peracarida, as does the most recent treatment (Hessler, 1999) of the order. Gutu and Iliffe (1998) described a new (third) species of hirsutiid from anchialine and submarine caves in the Bahamas and suggested that the family be removed to a new order, the Bochusacea (separate order status for the hirsutiids had been suggested also by Sanders et al., 1985). The other family of Mictacea (Mictocarididae) was then proposed by Gutu (1998) to belong to a new order, Cosinzeneacea, which would include as suborders the Spelaeogriphacea and Mictacea. We have not followed the suggestions of Gutu and Iliffe (1998) and Gutu (1998).

## ORDER AMPHIPODA

The Amphipoda, despite a large number of dedicated workers and numerous proposed phylogenies and classificatory schemes, remain to a large extent an unresolved mess. Families proposed by one worker often are not recognized by another, and disparate classifications based on poorly defined features seem to be the rule. The Gammaridea, containing the vast majority of amphipod families, is the most confusing suborder, although several workers (e.g., Kim and Kim, 1993) have proposed cladistically based rearrangements of the taxa. We should comment especially on the "semi-phyletic classification" put forth by Bousfield and Shih (1994) in the journal *Amphipacifica*. This classification apparently is being used as the basis for amphipod classification in an upcoming publication on common names of North American invertebrates overseen by the American Fisheries Society (although "minor changes may yet be made"; E. Bousfield, pers. comm., March, 1999). Consequently, the Bousfield and Shih (1994) classification or its successor in the AFS publication (see Bousfield, 2001) is likely to be cited often in the years to come. Although the Bousfield and Shih (1994) work is of value in reviewing previous classificatory attempts in recent years, we have not adopted it here. The classification divides the group into the Amphipoda "Natantia" and Amphipoda "Reptantia," without assigning taxonomic rank to these divisions, and then lists the amphipod families under superfamily headings. Unfortunately, no authors or dates are provided for any of the higher taxa. A further point of frustration is that the authors include in that paper several different phylogenetic hypotheses based on different morphological features; however, the phylogenies are not concordant, so it is difficult to determine the characters on which they base their resulting "semi-phyletic" classification. These disparaging comments should not be taken as reflecting adversely on other papers from these authors. And indeed, a large number of papers in which various gammaridean amphipod superfamilies and families are revised have been authored by Bousfield and his colleagues in recent years and should be consulted by workers interested in those families. These works include Jarett and Bousfield (1994a, b, superfamily Phoxocephaloidea: Phoxocephalidae), Bousfield and Hendrycks (1994, superfamily Leucothoidea: Pleustidae; 1997, superfamily Eusiroidea: Calliopidae), Bousfield and Kendall (1994, superfamily Dexaminoidea: Atylidae, Dexaminidae), Bousfield and Hoover (1995, superfamily Pontoporeioidea: Haustoriidae), Bousfield and Hendrycks (1997, superfamily Eusiroidea: Calliopiidae), and Bousfield and Hoover (1997, superfamily Corophioidea: Corophiidae), and other papers in the journal *Amphipacifica*.

Following the Fourth International Crustacean Congress in Amsterdam, there was a meeting of amphipod specialists in Kronenburg, Germany (the IXth International Meeting on Amphipoda, July, 1998). One topic discussed in Kronenburg was "Whither amphipod family-level taxonomy?" The report stemming from that discussion (Vader et al., 1998) is interesting and informative, and we quote from it here:

> Currently the classification of the Amphipoda is still in a state of flux; the schedules of Jerry Barnard and Ed Bousfield, often not very compatible and neither of them based on cladistic analyses, are still prevalent. Discussions revolved around the bush-like evolution of the Amphipoda and envious comparisons to the Isopoda where the general classification appears clearer. Not unexpectedly, the classification problems of the Amphipoda were not solved! However, it was suggested that a cladistic analysis of the amphipod families should have high priority, simply to give a general idea of the overall relationships, and to generate topics for further studies.

To summarize, in the words of Les Watling (pers. comm.), "most of us working in the amphipod world would rather that the [gammaridean] families be listed alphabetically rather than by superfamilies."

Thus, somewhat to our disappointment, we have followed that group's suggestion and also the work of Barnard and Karaman (1991) (which has been followed by several other workers such as De Broyer and Jazdzewski, 1993) in listing alphabetically the many families of gammaridean amphipods in the current classification. This was done in the Bowman and Abele classification as well. The most recent treatment, an indispensable review by Bellan-Santini (1999), also lists the families of gammaridean amphipods (67 of them) alphabetically (in addition to listing another 24 families of questionable standing) without using superfamilies. This work (Bellan-Santini, 1999) differs from our compilation slightly and should be consulted by any serious student of gammaridean amphipods.

The alphabetical list of families presented here has the advantage of not espousing one worker's view over another (although because Barnard and Karaman, 1991, also listed families alphabetically, it could be argued that we are preferring their approach; E. Bousfield, pers. comm.). It has the additional advantage of signaling to future workers that the gammarideans are in serious need of further attention. However, our alphabetical listing has the clear disadvantage of discarding some groupings (e.g., corophioids, talitroids, lysianassoids) that seem to be fairly well accepted. An additional problem that should be noted is that, while we are avoiding superfamilies because they are controversial and/or not widely used, the same could be said for a large percentage of the families that we have chosen to recognize.

Works appearing subsequent to the Bowman and Abele (1982) classification that employ these superfamily groupings (although not all in perfect agreement as to the constituent families) of the gammarideans include Schram (1986), Ishimaru (1994), Bousfield (1983), and Bousfield and Shih (1994). These papers should be consulted for further information on gammaridean superfamily hypotheses. Further advances in our understanding of amphipod phylogeny were presented as part of the 10th Colloquium on Amphipoda (Heraklion, Crete, April, 2000) and include Berge et al. (2000), Bousfield (2000a, b), Serejo (2000), and Lowry and Myers (2000), abstracts of all of which are available via the Amphipod Homepage hosted by Old Dominion University in Norfolk, Virginia (URL http://www.odu.edu/%7Ejrh100f/amphome).

## SUBORDER GAMMARIDEA

Gammaridean amphipod families that have been described or recognized since the Bowman and Abele (1982) list include, in alphabetical order of the families, Acanthonotozomellidae (by Coleman and Barnard, 1991), Amathillopsidae (recognized by Coleman and Barnard, 1991, credited to Pirlot, 1934, but considered only a subfamily of the Epimeriidae by Lowry and Myers, 2000), Allocrangonyctidae (by Holsinger, 1989), Aristiidae (by Lowry and Stoddart, 1997), Bolttsiidae, Cardenioidae, Clarenciidae (all by Barnard and Karaman, 1987), Cheidae (by Thurston, 1982), Condukiidae (by Barnard and Drummond, 1982), Cyphocarididae (by Lowry and Stoddart, 1997), Dikwidae (by Coleman and Barnard, 1991, suggested to be only a tribe within the subfamily Amathillopsinae by Lowry and Myers, 2000), Didymocheliidae (by Bellan-Santini and Ledoyer, 1986), Endevouridae (by Lowry and Stoddart, 1997), Ipanemidae and Megaluropidae (by Barnard and Thomas, 1988), Metacrangonyctidae (by Boutin and Missouli, 1988), Micruropidae (by Kamaltynov, 1999), Odiidae (by Coleman and Barnard, 1991, but see Berge et al., 1998, 1999, who believe that the Odiidae is paraphyletic and that its genera belong instead within the Ochlesidae), Opisidae (by Lowry and Stoddart, 1995), Pachyschesidae (by Kamaltynov, 1999), Paracalliopiidae (by Barnard and Karaman, 1982), Paracrangonyctidae (by Bousfield, 1982), Paraleptamphopidae (by Bousfield, 1983), Perthiidae (by Williams and Barnard, 1988), Phoxocephalopsidae (by Barnard and Clark, 1984, who credit Barnard and Drummond, 1982), Phreatogammaridae (by Bousfield, 1982), Pseudamphilochidae Schellenberg (revised and reinserted by Barnard and Karaman, 1982), Podoprionidae (by Lowry and Stoddart, 1996), Pseudocrangonyctidae (by Holsinger, 1989), Scopelocheiridae (by Lowry and Stoddart, 1997), Sinurothoidae (by Ren, 1999), Sternophysingidae (by Holsinger, 1992), Urohaustoriidae (by Barnard and Drummond, 1982), Valettidae (by Thurston, 1989), Wandinidae (by Lowry and Stoddart, 1990), and Zobrachoidae (by Barnard and Drummond, 1982). Additionally, we include the Podoceridae Leach, as this appears to be a widely recognized and relatively uncontroversial family (e.g., in Barnard and Karaman, 1991, and Bellan-Santini, 1999), although it was not listed by Bowman and Abele (1982). Iphimedioid amphipods, like many other groupings, are currently being revised, and as a result, some of the names and ranks above will undoubtedly change (see Lowry and Myers, 2000).

The family Lepechinellidae Schelenberg, listed in Bowman and Abele (1982), has been removed. Barnard and Karaman (1991) listed the genus *Lepichenella* in the Dexaminidae and considered the Lepichenellidae a synonym of the Dexaminidae (but note that Bousfield and Kendall, 1994, treated the Lepichinellidae as a subfamily of the Atylidae). The family Conicostomatidae is listed in the Zoological Record (1983, vol. 20, section 10), where it is attributed to Lowry and Stoddart (1983). However, although those authors recognized it as a grouping of related taxa, they did not establish it as a family in their 1983 paper, and they have not done so subsequently (J. Lowry, pers. comm.). Thus, the listing of the family in the Zoological Record is in error. The family Anamixidae is maintained in our classification, although there is reason to believe that this family was erected to accommodate what are turning out to be highly derived males of some species of the Leucothoidae (J. Lowry, pers. comm.). If true, the Anamixidae will have to be synonymized at some point. A few workers asked us to "correct" the spelling of the family name Liljeborgiidae to Lilljeborgiidae to reflect the fact that the family name honors William Lilljeborg (1816–1908). The confusion stems from the fact that Vilhelm Liljeborg changed the spelling of his name to William Lilljeborg sometime in the early 1860s. When Bate (1862) established the genus *Liljeborgia,* he used the then-correct spelling honoring Vilhelm Liljeborg. Thus, when Stebbing in 1899 established the family Lilejborgiidae based on the genus *Liljeborgia,* he was obliged to use this spelling as well even though, by that time, the man was known as William Lilljeborg (J. Lowry, pers. comm., and see Vader, 1972). (As an aside, the spelling of the genus *Lilljeborgiella,* erected by Schellenberg in 1931, is therefore also correct, as by that time the name was William Lilljeborg).

All of the 67 families that Bellan-Santini (1999) lists as those that "ne présent pas actuellement de problème majeur d'interprétation" are included in our list. Bellan-Santini (1999) also lists another 24 families that do present problems, and some of those are in our list as well. Some of the names and dates attributed to some families differ between our list and hers as well.

## SUBORDER CAPRELLIDEA

Takeuchi (1993) indicated that the Caprellidea may not be monophyletic but stopped short of proposing a new classification of the group. His results (Takeuchi, 1993, figs. 1, 5) indicated that the phtisicids are the sister group to all other caprellideans and that the paracercopids are more closely related to the caprellid-caprogammarid line (he did not deal with the parasitic family Cyamidae). Thus, we have removed the family Paracercopidae from the superfamily Phtisicoidea and have placed it instead in the superfamily Caprelloidea, leaving the Phtisicidae the sole family of the Phtisicoidea. We saw this move as preferable to creating yet another superfamily (to contain the paracercopids) in an already taxon-dense suborder. In the same year and in the same volume, Laubitz (1993) described two new caprellidean families (Caprellinoididae and Pariambidae). She also recognized as valid the Protellidae McCain and tentatively suggested some evolutionary lines or trends within and leading up to the Caprellidea. Some of these ideas differ from those proposed by Takeuchi (1993), although both workers recognize the same eight families (as does Bellan-Santini, 1999). Also in that same volume, Kim and Kim (1993) suggested affinities between caprellideans and corophioids. Margolis et al. (2000) have suggested that the Cyamidae may be closer to the Caprogammaridae-Caprellidae lineage rather than to the Caprellinoididae-Phtiscidae line, as suggested by Laubitz (1993). Several names and dates have reverted to earlier workers (suggestions of L. Holthuis, pers. comm.). The families Aeginellidae and Dodecadidae have been deleted, as they are now considered subfamilies of the Caprellidae and Phtisicidae (K. Larsen, pers. comm.; Laubitz, 1993). See also Bellan-Santini (1999).

## SUBORDER HYPERIIDEA

Workers familiar with hyperiideans may wonder why we did not follow the revision of the Hyperiidea by Vinogradov et al. (1982, with English translation edited by D. Siegel-Causey appearing in 1996). While that work contains much updated information concerning the biology of hyperiideans and nomenclatural changes below the level of family, the authors followed, for the higher classification, the earlier work by Bowman and Gruner (1973). Thus, the Bowman and Abele (1982) classification is the more current of the two for higher level taxa, although workers will want to consult the Vinogradov et al. volume for information within families and genera (D. Causey, pers. comm.). Our classification is also consistent with the classifications of Schram (1986, which in turn was based largely on Bousfield, 1983) and Bellan-Santini (1999). Kim and Kim (1993) suggested that hyperiids may be related to certain leucothoid members (Amphilochidae and Stenothoidae) of the Gammaridea.

## SUBORDER INGOLFIELLIDEA

Several workers (e.g., J. Holsinger, pers. comm.) have pointed out that the ingolfiellids and metaingolfiellids may not justify their own suborder and could probably be accommodated within the Gammaridea. Indeed, Bowman and Abele (1982) listed them alphabetically among the other gammaridean families. However, Holsinger notes at the same time that this view is not universally shared by other amphipod workers, and most workers (e.g., Bellan-Santini, 1999) continue to treat these two families as the sole members of the suborder Ingolfiellidea. Vonk and Schram (1998) argue for maintaining separate status for the group. We have retained their separate status pending further investigations into the group's affinities.

## ORDER ISOPODA

The diversity of and fascination with isopods are reflected in the relatively large number of carcinologists currently working on isopod systematics and phylogeny. Although it is encouraging to see so many skilled workers dedicated to resolving questions of isopod systematics, there are negative aspects, one of which is the relatively large number of responses we received that contained conflicting ideas or information. For the most part, we have relied on the rather straightforward list of the marine isopods that has been posted on the World Wide Web by B. Kensley and M. Schotte (http://www.nmnh.si.edu/iz/isopod). However, in that compilation, the various suborders and their constituent superfamilies and families are arranged alphabetically. Brusca and Wilson (1991), while proposing some phylogenetic changes that would seriously alter the arrangement of groups as presented here (and at the same time countering several of the hypotheses forwarded earlier by Wägele, 1989), stopped short of proposing a new classification based on their hypothesis. Their feeling was that insufficient evidence had been amassed for proposing classifications based on the phylogenetic hypotheses they were presenting as testable ideas. The Brusca and Wilson (1991) analysis was criticized by Wägele (1994), who in fact used their paper to point out potential pitfalls in any attempt at computer-generated cladistic analyses. Wägele (1994) was in turn rebutted by Wilson (1996), who was answered by Wägele (1996), and it would seem that we have a long way to go before any consensus concerning isopod phylogeny (not to mention phylogenetic method) is reached. Thus, our classification is in some ways a step backward in that we continue to recognize some groups, such as the Flabellifera, that appear clearly paraphyletic (following the analyses of both Brusca and Wilson, 1991, and Wägele, 1989) but for which no alternative classifications have been proposed. In the most recent overall treatment of isopods, Roman and Dalens (1999) continue to recognize the Flabellifera as well while acknowledging that it is a heterogeneous assemblage.

Additionally, many changes, especially those concerning names and dates of the authorities credited with establishing families but also concerning whether or not to recognize a particular family, have been incorporated at the request of some of the major workers (e.g., L. Holthuis, B. Kensley, R. Brusca, G. Poore, W. Wägele, and G. Wilson) via personal communications. It has not always been possible for us to verify these suggestions. Often, despite a rather large library on crustacean systematics at our disposal, we have been unable to see the original references. In cases where we received conflicting information (such as whether the family Arcturidae should be credited to White, 1850 vs. Bate and Westwood, 1868 vs. Sars, 1899) and/or we could not verify by checking on all of the suggested references ourselves, we have chosen the first known usage (in this case, using Arcturidae White, 1850, which turns out to be correct according to G. Poore, who owns the book) in accordance with ICZN article 50.3.1. One such change involves the establishment of a large number of families and superfamilies credited to Latreille. L. Holthuis (pers. comm.) assures us that 1802 is the correct date for the many taxa that have been, in the past, credited to Latrielle (1803) (see earlier section on names, dates, and the ICZN).

Major papers suggesting changes in how we organize the Isopoda that have appeared subsequent to Bowman and Abele (1982) include Wägele (1989) and Brusca and Wilson (1991). Poore (2001a) presented a phylogeny of the Anthuridea suggesting relationships among the six families (two new), but to our knowledge, there have not as yet been names proposed for the divisions suggested by him. The most recent review, by Roman and Dalens (1999), recognizes eight suborders. Their arrangement differs from ours in that (1) they recognize the suborder Gnathiidea, which we do not, and (2) they do not recognize the suborders Microcerberidea and Calabozoidea, which we do, for reasons discussed below.

Concerning the former suborder Gnathiidea, Brusca and Wilson (1991) suggested that the gnathiids were derived from among the families traditionally thought of as "flabelliferan" isopods (a group that they demonstrate is not monophyletic). Wägele (1989, pers. comm.) also would remove the gnathiids from their own suborder, but his preference was to place them among the Cymothoida, a group he recognizes as containing a large number of former Flabellifera families. We have, for the current classification, removed the gnathiids from their own superfamily and have placed them within the Flabellifera, knowing that the Flabellifera itself is not monophyletic and must some day be extensively revised. L. Holthuis (pers. comm.) has suggested that we credit the family name Gnathiidae to Leach (1814) rather than to Harger (1880), as was used by Bowman and Abele (1982) and Roman and Dalens (1999).

## SUBORDER PHREATOICIDEA

Wilson (pers. comm.) suggests that many of the subfamilies of the Amphisopodidae recognized by Nicholls (1943, 1944) will need to be elevated to family level (e.g., as Hypsimetopodidae, Mesamphisopodidae, Phreatoicopsididae) once this suborder is revised (see also Wilson and Johnson, 1999; Wilson and Keable, 1999, 2001). Our classification follows Roman and Dalens (1999) in recognizing three families (the same three that appear in Bowman and Abele, 1982). By listing the phreatoicids first among all isopod suborders, we are acknowledging the primitive nature of these isopods. Brusca and Wilson (1991) and Wilson and Johnson (1999) have indicated that the phreatoicideans, all of which are restricted to Gondwanan fresh waters, may be "the earliest derived isopod Crustaca" (Wilson and Johnson, 1999:264). The phreatoicidean fossil record extends back to the Carboniferous (Wilson and Johnson, 1999).

## SUBORDER ANTHURIDEA

Within this suborder, the family Antheluridae was described by Poore and Lew Ton (1988) and the families Expanathuridae and Leptanthuridae were described recently by Poore (2001a; see also Poore, 1998). Our treatment differs from that of Roman and Dalens (1999) in that we include six families. Roman and Dalens do not recognize the family Antheluridae and of course could not have known about the Expanathuridae and Leptanthuridae.

## SUBORDER MICROCERBERIDEA

Wägele (1983) placed the family Microcerberidae within the Aselloidea; Brusca and Wilson (1991) considered the Microcerberoidea the sister group to the Asellota and consequently suggested they not be included among the Asellota. Our treatment of the family as belonging to its own suborder and superfamily follows Bowman and Abele (1982) but is also in keeping with the suggestion of Brusca and Wilson (1991). Additionally, we now treat the monotypic family Atlantasellidae in this suborder on the recommendation of G. D. F. Wilson (pers. comm.).

## SUBORDER FLABELLIFERA

Brusca and Wilson (1991) showed that the Flabellifera was a paraphyletic grouping, a finding that has been suggested also by other workers. Wägele (1989) (rebutted to some degree by Wilson, 1996) argued for dividing the flabelliferan families into two somewhat smaller groups, the Cymothoida and Sphaeromatidea (see Wägele, 1989). Wägele would remove from the Flabellifera the family Atlantasellidae (which he considers an Aselloidea). The families Aegidae, Anuropidae, Argathonidae, Cirolanidae, Corallanidae, Cymothoidae, and Tridentellidae would belong to his grouping Cymothoida Leach, 1814. The remaining families (Bathynataliidae, Hadromastacidae, Keuyphyliidae, Limnoriidae, Phoratopodidae, Plakarthriidae, Serolidae, Sphaeromatidae, and Tecticepitidae) he would place in the Sphaeromatoidea. Thus, the two most current and most ambitious schemes of isopod phylogeny, although agreeing in some respects, do not agree even closely on how to treat the former flabelliferan families (see also Brandt et al., 1999, for a comparison of phylogenetic hypotheses of sphaeromatoid families in light of the fossil family Schweglerellidae). Roman and Dalens (1999) recognize the Flabellifera, and divide it into three superfamilies: Cirolanoidea (seven families), Sphaeromatoidea (two families), and Seroloidea (two families). We have retained the Flabellifera for the current classification, knowing that this assemblage cannot be considered monophyletic, and for now, we have avoided the use of superfamilies. Recent fossil finds (see Brandt et al., 1999) have pushed back the origin of some former flabelliferan isopods, indicating that the sphaeromatoid isopods, at least, are of Late Jurassic ancestry or older.

Within the Flabellifera, the following changes have been incorporated (listed alphabetically by family): Ancinidae (elevated to family status by N. L. Bruce, 1993), Argathonidae (removed per R. Brusca, pers. comm.), Bathynomidae (removed per B. Kensley, pers. comm.), Excorallanidae (removed per B. Kensley, pers. comm.), Hadromastacidae (described by Bruce and Müller, 1991), Lynseiidae (described by Poore, 1987; removed per Cookson and Poore, 1994; see also Bruce, 1988), Protognathiidae (described by Wägele and Brandt, 1988; moved from Gnathiidea per R. Brusca and also G. Wilson, pers. comm.), Tecticepitidae (originally described as a subfamily by Iverson, 1982; elevated to family status by N. L. Bruce, 1993), and Tridentellidae (described by Bruce, 1984).

N. L. Bruce (1993) presented a key to the known flabelliferan families, reappraised the family Sphaeromatidae Latreille (a family in rather dire need of internal revision; see Harrison and Ellis, 1991), and recognized as families the Ancinidae Dana and Tecticipitidae Iverson.

G. Poore (pers. comm.) informs us that the Aegidae is correctly attributed to White (1850) rather than to Leach (there are no families mentioned in the only paper that Leach published in 1815, the date given in Bowman and Abele for this family). He also informs us that the families Ancinidae, Cirolanidae, and Serolidae are correctly attributed to Dana (1852) instead of 1853 (as in Bowman and Abele, 1982).

Bowman and Abele (1982) used the spelling Anuropodidae for this isopod family, while noting (1982: 21) that the tanaid family Anuropodidae Băcescu was a homonym of the isopod family Anuropodidae Stebbing. ICZN Opinion 1357 (ICZN, 1985b) dictated that the spelling of the isopod family should be Anuropidae to remove the homonymy, and thus we use Anuropidae as the correct spelling of this isopod family.

The Plakarthriidae Hansen is, according to G. Poore (pers. comm.), "an effective replacement name for Chelonidiidae Pfeffer, 1887, but is conserved under ICZN article 40"; Dr. Poore suggests that the date 1887 should follow Hansen, 1905, in parentheses, as Plakarthriidae Hansen, 1905 (1887).

## SUBORDER ASELLOTA

According to G. Wilson and G. Poore (pers. comm.), the currently recognized superfamilies of the Asellota are either poly- or paraphyletic (see also Wilson, 1987) and will not stand the test of time. Roman and Dalens (1999) treat the Asellota as being comprised of four superfamilies (down one from Bowman and Abele, 1982; the Protallocoxoidea and its single family, Protallocoxidae, have been removed). We have followed this arrangement here, recognizing the superfamilies Aselloidea, Stenetrioidea, Janiroidea, and Gnathostenetroidea. The superfamily Pseudojaniroidea, proposed by Wilson (1986), has been removed at his suggestion (G. Wilson, pers. comm.; see also Serov and Wilson, 1999). Its former family, the Pseudojaniridae, has been transferred to the Stenetrioidea following the revision of the Pseudojaniridae by Serov and Wilson (1999).

In the superfamily Aselloidea, the family Atlantasellidae has been removed. Brusca and Wilson (1991) suggested its removal to the Microcerberoidea, where we have placed it. Although Roman and Dalens (1999) treat the family Microcerberidae as a member of the Aselloidea, we are keeping it in its own suborder (Microcerberidea) and superfamily (Microcerberoidea) as per Bowman and Abele (1982) (as noted earlier). Thus, the Aselloidea presently contains only the Asellidae and Stenasellidae.

The superfamily Stenetrioidea now contains the Pseudojaniridae (as noted above), although Roman and Dalens (1999) have kept it at one family, the Stenetriidae.

Within the enormous superfamily Janiroidea, the Abyssianiridae was removed (incorporated into the Paramunnidae) following Just (1990). Species formerly within that family are now considered to belong to the Paramunnidae. The former families Eurycopidae, Ilyarachnidae, and Munnopsididae are now considered subfamilies of the Munnopsididae (Wilson, 1989). The Microparasellidae is apparently polyphyletic; "some taxa may be moved to the Vermectiadidae or put in a new family; *Microparasellus* will stay in the Janiroidea" (Wilson, pers. comm.). The Janiridae was shown to be nonmonophyletic by Wilson (1994) but remains a valid family; some of its genera will eventually be reassigned to other families. The Katianiridae was described by Svavarsson (1987). Although the family Pleurogoniidae is recognized by some workers (e.g., Roman and Dalens, 1999), we have removed it at the suggestion that it is a junior synonym of the Paramunnidae (G. Poore, pers. comm.; G. Wilson, pers. comm.). The family Pseudomesidae was sunk into the Desmosomatidae by Svavarsson (1984). Although the family Santiidae is credited to Kussakin (1988) by many workers (e.g., Wolff, 1989), it was first used (in a figure) by Wilson (1987). In Wilson's (1987) paper, he acknowledges Fresi et al. (1980) as the source for one of the phylogenetic trees in that paper (Wilson's fig. 5B). However, Fresi et al. (1980) did not include the Santiidae in their figure; it was apparently added (and therefore first used) by Wilson (1987). Thus, we have credited the family Santiidae to Wilson. Cohen (1998), in his review of the family Dendrotiidae, explains why this spelling of the family name is preferred over Dendrotionidae (used by Lincoln and Boxshall, 1983). Interested workers should also consult Roman and Dalens (1999), whose list of families differs from ours in several respects.

The superfamily Protallocoxoidea and family Protallocoxidae were removed per G. Wilson (pers. comm.).

The superfamily Gnathostenetroidoidea contains the families Gnathostenetroididae and Protojaniridae (following Roman and Dalens, 1999). Additionally, the interesting family Vermectiadidae was described by Just and Poore (1992), and our tentative inclusion of the vermectiadids in the superfamily Gnathostenetroidoidea is based mostly on the recommendation of R. Brusca (pers. comm.).

## SUBORDER CALABAZOIDA

This family (Calabozoidae) and its suborder were erected by Van Lieshout (1983). Brusca and Wilson (1991) suggest that the calabazoids are oniscideans and so they should probably be moved, but we have not done so in this classification. Wägele (pers. comm.) points out that the ending -oidea should be reserved for superfamilies and suggested that we change the spelling of the suborder to Calabazoida, which we have done.

## SUBORDER VALVIFERA

Within the Valvifera, several families have been added since the Bowman and Abele (1982) classification. The family Austrarcturellidae was described by Poore and Bardsley (1992), and the families Antarcturidae, Arcturididae, and Rectarcturidae were added by Poore (2001b). Poore (2001b) also recognized the Holidoteidae, crediting it to Wägele (1989), who first suggested it as a subfamily. Current research shows that the family Amesopodidae is probably a junior synonym of the Arcturidae (G. Poore, G. Wilson, pers. comm.), and so we have removed it, although the family was listed by Roman and Dalens (1999), who did not list the Austrarcturellidae. Thus, we recognize 11 families, 4 more than did Bowman and Abele (1982). The family Arcturidae, credited by Bowman and Abele (1982) to Sars, is correctly credited to Dana (1849), and the family Idoteidae is correctly attributed to Samouelle (G. Poore, pers. comm.).

## SUBORDER EPICARIDEA

Wägele (1989, pers. comm.) suggested that all of the epicaridean families we have listed should be treated as families or subfamilies of the Cymothoida Leach (see above). We have not made this rather radical change and instead have followed the more conservative classification given by Trilles (1999). Trilles (1999) divides the epicaridean families into two sections, Bopyrina and Cryptoniscina, which we have treated as superfamilies (Bopyroidea and Cryptoniscoidea) to allow a more consistent spelling and in keeping with our treatments of other peracarid groups. In the Bopyroidea are the three families Bopyridae, Dajidae, and Entoniscidae (all of which were listed by Bowman and Abele, 1982). In the section (now superfamily) Cryptoniscoidea, Trilles (1999) treats an additional eight families not listed by Bowman and Abele (1982); the family Liriopsidae has been deleted (see arguments in Grygier and Bowman, 1990, 1991; Trilles, 1999). Thus, 11 epicaridean families are recognized. The families added since Bowman and Abele (1982) are not newly described families but instead represent recognition of formerly described families that were treated in the past, at least by some authors, as subfamilies of the Cryptoniscidae, for which Bowman and Abele (1982), followed by Schram (1986), used the name Liriopsidae (see Grygier and Bowman, 1990). Crediting authorship of the family Cryptoniscidae (and thus Cryptoniscoidea) to Kossman rather than to Gerstaecker is based on the correction published by Grygier and Bowman (1991). Following Trilles (1999), we also do not recognize the family Microniscidae Müller for the genus *Microniscus,* although this family is still listed in some compendia (e.g., by Brasil-Lima, 1998:641, in Young, 1998). The spelling Cabiropsidae used by Trilles (1999) and some earlier workers is corrected to Cabiropidae based on the explanation given by Sassaman (1992).

## SUBORDER ONISCIDEA

The relationships of the terrestrial isopod groups to one another and to marine relatives are still poorly understood. Although Schmalfuss (1989, in Ferrara, 1989) proposed some relationships among oniscideans and compared the classification of oniscideans presented by Holdich et al. (1984) with a new one based on his analysis, Schmalfuss' work was based on relatively few characters and was criticized by Brusca (1990). Wägele (pers. comm.) informs us that there are "enormous advances that will be published next year" concerning the phylogeny of the Oniscidea and that several groups presented here are not monophyletic; further, he informs us that the "section" Diplochaeta is currently being revised. Until these advances become known to us, we are unsure as to what relationships our classification should suggest. Holdich et al. (1984) used two infraorders (the Tylidae were placed in a separate infraorder, Tylomorpha), and within the infraorder Ligiamorpha they recognized three sections. Schmalfuss (1989) did not employ the infraorder level and instead divided all oniscideans among four major sections. More recent arrangements of the oniscidean families have been proposed by Erhard (1995) and Tabacaru and Danielopol (1996a, b; see also Roman and Dalens, 1999, who followed mostly Schmalfuss, 1989, and also Mattern and Schlegel, 2001). Many workers (e.g., Souza-Kury, 1998, in Young, 1998) list the oniscidean families alphabetically.

We have maintained the two-infraorder system and have not recognized the new section Microchaeta proposed by Schmalfuss. The four families Helelidae, Irmaosidae, Pseudarmadillidae, and Scleropactidae have been removed from any infraorder or superfamily, as their status is indeterminate (R. Brusca, pers. comm.). For the currently accepted family names (as well as authors and dates, which were not included by Schmalfuss), we have had to rely primarily on the alphabetical list of oniscidean families maintained on the Smithsonian's server (Kensley et al., 1998; URL http://www/nmnh.si.edu/iz/isopod), which is based on Schmalfuss' families (the terrestrial isopod list is also accessible via the Kensley et al. list of marine isopods, URL gopher://nmnhgoph.si.edu:70/11/.invertebrate/.crustaceans). Users of the terrestrial isopod list are strongly cautioned by the authors (Kensley et al., 1998):

> This list is thus intended as a rough guide to the astounding array of names and taxa in the Oniscidea. Synonymy will be rampant in the list. We have tried to use the most current interpretations of some genera and families. Nevertheless, we realise that in no way do we even begin to resolve the taxonomic confusion that reigns in this group. There is uncertainty regarding the familial placement of some genera, and there will certainly be repetition of the same specific name under different genera. There are omissions from the list, either of names of taxa that we've completely missed, or of authors and dates of publication and/or of localities that we have been unable to find.

We are aware of only two newly described oniscidean families since 1982: Ferrara and Taiti (1983) described the family Irmaosidae, and Schultz (1995) described the Dubioniscidae (see Souza-Kury, in Young, 1998:656). Establishment of the family Platyarthridae is credited to Verhoeff (rather than to Vandel) by Ferrara and Taiti (1989), who also note that the families Bathytropidae and the Platyarthridae might coincide. G. Poore (pers. comm.) notes that the Styloniscidae Vandel, 1952, is a replacement name for the Patagoniscidae Verhoeff, 1939, and is conserved under ICZN article 40; he therefore recommends that the earlier date appear in parentheses, as Styloniscidae Vandel, 1952 (1939). Characters that define the various groupings of the oniscideans are given by Roman and Dalens (1999), although workers should note that the characters and groupings based on them are, in some cases, not universally accepted. A recent molecular analysis (Mattern and Schlegel, 2001) based on ssu rDNA suggests that Crinochaeta and Synochaeta are monophyletic, and that these groups together are the sister taxon to the Diplochaeta.

## ORDER TANAIDACEA

Many of the major taxonomic changes suggested by the late J. Sieg were made prior to 1982 and were therefore incorporated into the Bowman and Abele classification. Subsequent to 1982, there were also some large-scale rearrangements suggested by Sieg (1983a, b, 1984, 1986a, b), but there has been almost no work done at higher levels of tanaid systematics since that time. Unfortunately, it now appears that many of the characters established or used by Sieg do not hold up well under scrutiny (see Larsen and Wilson, 1998), and it is not clear how many of Sieg's characters or numerous classificatory assignments will survive. Kim Larsen (pers. comm.) is actively studying the group and has kindly updated us, as far as is possible pending a thorough revision of the group. Additionally, he has provided us with many suggested changes. An excellent and comprehensive web site maintained by

Richard Heard and Gary Anderson now exists at URL http://tidepool.st.usm.edu/tanaids/index.html, and our arrangement of the group is the same as theirs.

Authorship of the Tanaidacea is now credited to Dana (1849) rather than to Hansen (1895) (L. Holthuis, pers. comm.). A review by M. Gutu and the late Jürgen Sieg (Gutu and Sieg, 1999) additionally includes fossil taxa (most of which were added by Schram et al., 1983). The classification in Gutu and Sieg (1999) differs from ours in that we include the family Tanapseudidae, not listed in Gutu and Sieg (1999), and in that we have deleted the Leptognathiidae (see below).

## SUBORDER TANAIDOMORPHA

The naturalness of the entire suborder Tanaidomorpha was questioned by Larsen and Wilson (1998), who noted that inconsistencies or contradictions in descriptions and illustrations of several authors "plague tanaidomorphan taxonomy." Larsen and Wilson also noted that several of Sieg's characters and subsequent classifications, which form the basis of our current understanding of tanaid systematics, have been found wanting. They conclude that "the current taxonomy . . . for the suborder Tanaidomorpha, heavily burdened by inconsistencies, is not useful at the present stage." It seems unlikely that the situation for the other suborders would be any better.

Within the Tanaidomorpha, the family Leptognathiidae was abandoned by Sieg (1986b) as it was found to be a junior synonym of Anarthruridae (Sieg, 1986b; see Larsen and Wilson, 1998). One of its constituent subfamilies was incorporated into the Anarthruridae Lang, and the other was elevated to familial rank (now the Typhlotanaidae Sieg). The family Agathotanaidae similarly was downgraded from a family to "tribe" status (Sieg, 1986b). Dates of establishment of the Nototanaidae and Pseudotanaidae (in the past, often credited to Sieg, 1973) have been changed from 1973 to 1976, as the 1973 work is an unpublished thesis that did not appear in published form until three years later (Sieg, 1976) (K. Larsen, pers. comm.).

Additional tanaidomorphan families described subsequent to the Bowman and Abele (1982) list include the Pseudozeuxidae and Typhlotanaidae, described by Sieg (1982) and Sieg (1986b), respectively.

## SUBORDERS NEOTANAIDOMORPHA AND APSEUDOMORPHA

Sieg (1983b) elevated to family status the Whiteleggiidae and placed within it the former family Leviapseudidae as a subfamily (Leviapseudinae) of the Whiteleggiidae. Sieg (1984) established the family Cyclopoapseudidae to accommodate a genus formerly in the Metapseudidae (Sieg, 1984; Larsen, pers. comm.), but the Cycloapseudidae is now considered a junior synonym of the Metapseudidae (Larsen, pers. comm.). The Parapseudidae was not accepted by Sieg (1986a, b) but has since been recognized as valid (see brief discussion in Gutu, 1996; K. Larsen, pers. comm.). See Gutu and Sieg (1999) for the most recent review.

## ORDER CUMACEA

Our classification follows the World Wide Web list compiled by Watling and Kornfield (URL http://nature.umesci.maine.edu/pub/Cumacea/data.html) as part of their National Science Foundation PEET training project. Their list is similar to that of Bowman and Abele, with two exceptions. First, the family Archaeocumatidae Băcescu, 1972, containing the single genus *Archaeocuma*, has been removed. Its establishment (in Băcescu, 1972) had been questioned earlier by Jones (1976), who felt that further confirmation was needed prior to accepting this family, and Watling (pers. comm.) informs us that this family is generally not recognized. However, the family is listed (considered valid) by Băcescu (1988) and by Băcescu and Petrescu (1999). Second, the family Gynodiastylidae Stebbing, 1912, has been included following Day (1980), Băcescu (1992), Băcescu and Petrescu (1999), and the above-mentioned web site. Thus, the total number of cumacean families remains at eight, as with the Bowman and Abele list, although the composition has changed. Watling (pers. comm.) also notes that the family Nannastacidae will very likely be split into two families in the near future.

Relationships within the Cumacea have been tentatively suggested recently by Haye and Kornfield (1999) on the basis of somewhat limited molecular data. Their suggestion is that those families with an articulated telson (families Bodotriidae, Leuconidae, and Nannastacidea) form a clade that is distinct from a second lineage containing the five families without an articulated telson. This grouping is not reflected in the current classification, where all families are instead listed alphabetically.

## SUPERORDER EUCARIDA

Most workers seem to be in agreement that the Eucarida is a valid (i.e., monophyletic) assemblage (but see arguments against a monophyletic Eucarida in Richter and Scholtz, in press). Schram (1984) noted that the eucarids "are destined for some kind of realignment," and he later (1986) apparently abandoned the group in his classification, treating euphausiids, amphionidaceans, and decapods as separate orders within the Eumalacostraca (Schram, 1986:543). Yet his cladogram (Schram, 1986:530) and his accompanying text (1986:529) depict the eucarid line as distinct, and he refers to the eucarids as one of the recognizable lines of eumalacostracan evolution. And indeed, most treatments consider the Eucarida a valid superorder of the subclass Eumalacostraca, as did Bowman and Abele (1982) and most treatments since then (e.g., Christoffersen, 1988; Ruppert and Barnes, 1994;

Brusca and Brusca, 1990; Mayrat and Saint Laurent, 1996; Camp, 1998 (in Camp et al., 1998); Young, 1998). But as with nearly all other crustacean assemblages, this grouping has its opponents as well. Most of the disagreement concerns whether the mysidaceans (i.e., either mysids, lophogastrids, or both) should be placed here (see earlier discussions on mysids and lophogastrids) and what the relationships are among the three currently recognized orders Euphausiacea, Amphionidacea, and Decapoda (see Jarman et al., 2000; Richter and Scholtz, in press). Eucarid relationships have been analyzed by Schram (1984) and by Christoffersen (1988). Our classification is consistent with both of these analyses at higher levels but differs in the constituent suborders.

### ORDER EUPHAUSIACEA

The Euphausiacea still contains only the two families Bentheuphausiidae (monotypic) and Euphausiidae (all other species). The treatment by Baker et al. (1990) follows this arrangement as well. A recent analysis of 28S rDNA sequence data by Jarman et al. (2000) suggests that euphausiaceans may be more closely related to the Mysida than to the Decapoda.

### ORDER AMPHIONIDACEA

This order remains monofamilial and monogeneric (*Amphionides*).

### ORDER DECAPODA

The decapods have been the subject of more published papers than have all other crustacean groups combined. This popularity stems in part from the economic importance of many groups (especially penaeoid shrimps, palinurid and nephropid lobsters, and portunid and xanthoid crabs) but also in part because of their marvelous diversity. The convenient size of most decapods predisposed them to become subjects of some of the earliest papers using biochemical and molecular data to resolve crustacean relationships. Yet we are as far from reaching a consensus on the relationships among the decapods as we are for the more obscure groups, and opinions and datasets remain sharply divided. In the treatment that follows, we have tried to address the many changes and arrangements that have been suggested since 1982 under the taxonomic heading for each major group of decapods. However, we are certain to have missed several important papers, and we hasten to remind the reader that the literature on this topic is vast. In general, we have settled on a fairly conservative classification of the decapods, knowing that, as with all other crustacean taxa, this group is destined for revision. Some of the many reviews of decapod classification that have appeared since the Bowman and Abele (1982) classification are Felgenhauer and Abele (1983), Abele and Felgenhauer (1986), Kim and Abele (1990), Abele (1991), Holthuis (1993a), and Scholtz and Richter (1995).

The creation of two major branches of decapods, Dendrobranchiata and Pleocyemata, by Martin Burkenroad (1963, 1981) was a rather bold departure from previous schemes of decapod classification. According to Fenner Chace (pers. comm.), T. Bowman more or less accepted Burkenroad's arguments without much questioning, and thus the use of the Dendrobranchiata and Pleocyemata in the Bowman and Abele (1982) classification. Chace (pers. comm.) feels that there is ample evidence for elevating many of the major groups of the Decapoda as Burkenroad did with the penaeoids and that singling out the penaeoid shrimp was to assign that group an artificial distinction. He is not alone. Holthuis (1993a; see especially pages 11–13 for a concise historical overview of the many attempts to classify the decapods) felt that treating the penaeoids as a separate group (the Dendrobranchiata) equal in rank to the combined Natantia + Macrura Reptantia + Anomura + Brachyura (the Pleocyemata of Burkenroad) was unsatisfactory. Holthuis (1993a) proposed to revert to the older classifications and treated the Natantia and the Macrura Reptantia as "full suborders of equal rank with the Anomura and Brachyura." In his own words (Holthuis, 1993a:6):

> I know that this classification will generally be considered old-fashioned: in several modern handbooks the suborder Natantia has been abandoned altogether; a small part of it, namely the Penaeoidea, is elevated to the rank of a separate suborder Dendrobranchiata while the rest of the Natantia plus the Macrura Reptantia, plus the Anomura, plus the Brachyura are placed in a single suborder Pleocyemata. This to me seems a very artificial and unsatisfactory arrangement, and I therefore still keep to the old classification.

This "old" classification to which he refers, probably because of its simplicity and relative lack of controversy, is often encountered in popular and lay versions of crustacean classification. As an example, the publishers of the BIOSIS and Zoological Record databases (see URL http://www.york.biosis.org/zrdocs for the BIOSIS/Zoological Record Taxonomic Hierarchy, Section 10, Crustacea) have "thrown up their hands in despair" (Chace, pers. comm.) and have reverted to this older and simpler classification. There, Natantia is treated as a taxon containing all of the known shrimp groups (Penaeidea, Caridea, and Stenopodidea) and the Reptantia is treated as containing the anomurans, astacurans, brachyurans, and palinurans.

Yet the distinct nature of the penaeoids (the Dendrobranchiata) has been supported by additional morphological (e.g., Schram, 1984, 1986), embryological, spermatological (e.g., see Jamieson, 1991a), and molecular data. Kim and Abele (1990) reviewed previous schemes of decapod classification and concluded, based on somewhat limited data from 18S rRNA, that the penaeids were distinct from other decapods. This view was supported with additional sequence data and additional taxon sampling by Abele (1991), whose review of morphological and molecular data supported a distinct Dendrobranchiata (the penaeoids) clade and also three other distinct clades corresponding to (1) the Caridea (including the procaridoids), (2) the Stenopodidea, and (3) a "reptant" lineage. (The latter lineage is responsible for most of the more troublesome remaining problems in decapod classification. As Abele (1991) stated, "there seems to be as many groupings of these taxa as there are authors who have studied them.") The artificiality of the "Natantia" is also pointed out by Christoffersen (1988a) and Scholtz and Richter (1995).

Thus, there is no morphological or molecular support for a natural "natantian" clade that contains all shrimp-like forms. The features that seem to unite the natantians appear to be primitive characters that do not clearly define a monophyletic group. Consequently, we have recognized the Dendrobranchiata and Pleocyemata on the basis of what appear (to us) to be shared, derived features of both morphological and molecular data.

Within the Dendrobranchiata, classification is relatively stable, mostly because there are relatively few taxa in this suborder. Relationships among the pleocyemate taxa are another story. If Caridea and Stenopodidea are treated as separate clades, then an argument could be made for recognition of the Reptantia (or Macrura Reptantia, following Holthuis, 1993a) as a natural taxon based on the work of Schram (1984, 1986), Abele (1991), Scholtz and Richter (1995), and others. Scholtz (pers. comm.) argues that the evidence for a monophyletic Reptantia is at least as convincing as the evidence for recognition of Caridea and other decapod infraorders, and we tend to agree. Yet the Reptantia of Abele (1991) and Scholtz and Richter (1995) differ as to the constituent groups, and we have opted for treating the "reptant" infraorders (Astacidea, Thalassinidea, Palinura, Anomura, and Brachyura) separately rather than combining them in a taxon that would be the sister group to the stenopodidean and/or caridean shrimps. Recognition of a natural "Reptantia" would involve using this name at the level of infraorder and then "demoting" the above five groups to just below the infraorder level, which would add considerably to the confusion in an assemblage that already contains a large number of taxonomic subdivisions.

Scholtz and Richter (1995) attempted to place the classification of the reptant decapods on firm cladistic footing. They argued (as did Christoffersen, 1988a) that the Reptantia was a clearly defined monophyletic taxon and that its sister group was possibly the Stenopodidea (which, according to other authors, are members of the same clade Pleocyemata). Thus, the branching sequence of the decapods would be Penaeoidea (Dendrobranchiata), then Caridea, Stenopodidea, and Reptantia; this much at least is consistent with other bodies of data (e.g., Schram, 1984, 1986; Jamieson, 1991a; Abele, 1991) (although Christoffersen (1988a:342) suggested that Stenopodidea was the sister group to a Caridea + Reptantia clade). In light of this support, it is curious, and possibly a mistake, that we have not included the Reptantia as a monophyletic clade in our classification, although inclusion or exclusion of the stenopodideans is unresolved. Scholtz and Richter (1995) argued convincingly for monophyly of some of the constituent reptant groups, such as the Brachyura and Anomura, but other arguments are (to us) less convincing. The Scholtz and Richter (1995) classification also included several new group names, such as the Achelata, Fractosternalia, Meiura, etc., which we feel are unlikely to persist (but note that some of these taxon names already have been employed (although not necessarily endorsed) in the papers of, e.g., Schmidt and Harzsch, 1999; Suzuki and McLay, 1998; Sternberg, 1996; Taylor et al., 1999; and Taylor and Schram, 1999). For reasons we feel are inappropriate for discussion in a review and compilation of this nature (mostly differences in how we would score certain morphological characters and the low number of specimens examined), we have not followed Scholtz and Richter here. In fairness, some of the characters proposed by Scholtz and Richter are well beyond our ability to comment on (such as the shape of thoracic and cephalic ganglia and the development of embryonic growth zones) and possibly provide fertile ground for further investigations. And we acknowledge and compliment them on an attempt to place decapod classification in a phylogenetic context, which our classification clearly does not do. But concerns raised by their questionable (to us) use of morphological characters caused sufficient doubt as to their overall scheme, and we have not accepted the Scholtz and Richter (1995) arrangement in the current classification.

The date of establishment of the name Decapoda has been changed to Latreille (1802) rather than Latreille (1803) (L. Holthuis, pers. comm.; see earlier comments in the section Names, Dates, and the ICZN).

## SUBORDER DENDROBRANCHIATA

Christoffersen (pers. comm.) would rather we employ the name Penaeidea Dana instead of Dendrobranchiata Bate, as the former name is older and "perfectly legitimate." Holthuis (pers. comm.) agrees but notes that "since Dendrobranchiata seems to [have] become generally accepted, I am quite willing to go along." Within the group, there have been no significant family-level or higher changes proposed (to our knowledge) since the Bowman and Abele (1982) classification. Authorship of the family Solenoceridae has been credited to Wood-Mason rather than to Wood-Mason and Alcock (Kensley, pers. comm.). Thus, our classification of the Dendrobranchiata is the same as that employed recently by Pérez Farfante and Kensley (1997).

## SUBORDER PLEOCYEMATA

The Pleocyemata contains all nonpenaeoid decapods, whether swimming (natant) or crawling (reptant). The group appears to be monophyletic based on morphological data (e.g., Schram, 1984, 1986; Scholtz and Richter, 1995) and molecular data (e.g., Kim and Abele, 1990; Abele, 1991).

## INFRAORDER STENOPODIDEA

For this section, we followed the classification provided by Holthuis (1993a), which does not appear to be very controversial. Authorship of the taxon Stenopodidea is changed from Bate (1888) to Claus (1872) at the recommendation of M. Tavares (pers. comm.). To our knowledge, there has been only one new family-level taxon described since the Bowman and Abele (1982) work. Schram (1986) erected the family Spongicolidae, so that there are now two recognized families of extant stenopodideans (see also Holthuis, 1993a). Schram et al. (2000) recently described the first known fossil stenopodidean, also attributed to the Spongicolidae.

## INFRAORDER CARIDEA

For the carideans, we followed, for the most part, the classification provided by Holthuis (1993a), which is very similar to that suggested by Chace (1992) (see also Vereshchaka, 1997b, for a key to caridean superfamilies modified slightly from Chace, 1992). But in contrast with the relative lack of controversy over dendrobranchiate or stenopodidean classification, there is apparently no consensus on the relationships or even the names of the incredibly diverse families of caridean shrimps. There have been several cladistic analyses conducted on groups of caridean families by M. Christoffersen (see especially Christoffersen, 1990). These studies would, if accepted, rearrange large numbers of caridean families. For example, in his 1986 paper, Christoffersen placed seven families (oplophorids, atyids, pasiphaeids, alvinocarids, bresiliids, psalidopodids, and disciadids) in the superfamily Atyoidea, in contrast with Chace (1992) and Holthuis (1993a), who treated the Atyoidea as containing only the family Atyidae. Christoffersen points out (pers. comm.) that, among the "glaringly non-monophyletic assemblages" in our current classification, are the Alpheoidea, Hippolytidae, Pandaloidea, and Nematocarcinoidea. Adding to Christoffersen's frustration (pers. comm.) is that, whereas many authors comment on the unsatisfactory state of current classifications, especially as concerns such "wastebasket" assemblages as the Hippolytidae and Pandaloidea, his own suggestions for novel arrangements have been slow to catch on. Chace (1997) recognizes the Hippolytidae Bate, and Holthuis (1993a) elected to synonymize a large number of Christoffersen's new taxa. Thus, we are left with the difficult task of following older yet clearly nonphylogenetic listings (e.g., Chace, 1992; Holthuis, 1993a) vs. cladistically generated phylogenetic arrangements (e.g., Christoffersen, 1987, 1988a, b, 1989a, b, 1990) that seem to have little following in the carcinological community and for which, in our estimation, some of the employed characters are questionable. We have followed Holthuis's lead, more in deference to his vast knowledge of the carideans than for any other reason, while acknowledging that there have been alternative phylogenetically based ideas presented in the literature. Only those superfamilies for which there have been changes subsequent to Bowman and Abele (1982) are mentioned below.

### Superfamily Galatheacaridoidea

The family Galatheacarididae and its superfamily Galatheacaridoidea were both described by Vereshchaka (1997b) for the species *Galatheacaris abyssalis* based on a single specimen. Additional specimens have since been found in the stomachs of deep-sea lancetfish (Chow et al., 2000).

### Superfamily Bresilioidea

This assemblage has long been recognized as being an artificial group in dire need of revision (e.g., see Forest, 1977). Holthuis (1993a) elected to treat the Bresiliidae as a family and placed in synonymy some of the recently proposed families (Agostocaridae, Alvinocarididae). We have treated the group as an (admittedly) artificial superfamily containing five caridean families that may or may not be related. Three of these families are new (i.e., they were not included in the Bowman and Abele (1992) classification): the family Agostocarididae was erected by Hart and Manning (1986), the Alvinocarididae was proposed by Christoffersen (1986), and the Mirocarididae was described by Vereshchaka (1997a).

Christoffersen's (1986) family Alvinocarididae is recognized to accommodate the majority of the morphologically similar "bresilioid" shrimp from hydrothermal vents. The family was more thoroughly (although still somewhat incompletely) diagnosed by Segonzac et al. (1993) in a footnote and also by Vereshchaka (1996, 1997a) (see also Shank et al., 1999). Vereshchaka (1997a) created a new genus (*Mirocaris*) and family, the Mirocarididae, for the hydrothermal vent shrimp described originally as *Chorocaris fortunata* by Martin and Christiansen (1995b).

### Superfamily Campylonotoidea

The family Bathypalaemonellidae was established (although without a description or diagnosis and without mention of the genus *Bathypalaemonella*; see Holthuis, 1993a:87) by Saint Laurent (1985). The family is placed in the superfamily Campylonotoidea on the recommendation of L. Holthuis (1993a:87, and pers. comm.).

### Superfamily Palaemonoidea

The family Euryrhynchidae Holthuis, 1950, was added on the recommendation of Holthuis (pers. comm.). The family Kakaducarididae was described by A. J. Bruce (1993) as a subfamily of the Palaeomonidae and is here treated as a family on the recommendation of L. Holthuis (pers. comm.).

### Superfamily Alpheoidea

Authorship of the family Ogyrididae remains credited to Holthuis (1955). Although Hay and Shore (1918) established the family Ogyridae, as noted by M. Tavares (pers. comm.), L. Holthuis (pers. comm.) points out that they based it on the type genus *Ogyris* Stimpson, 1860, which is a junior homonym of *Ogyris* Westwood and is thus invalid. Stebbing (1914) proposed the replacement genus *Ogyrides,* and thus the family name is Ogyrididae, first used as such by Holthuis (1955). We have not followed Christoffersen's (1987) suggestion to transfer the family Processidae to the Crangonoidea or to combine the alpheoids and crangonoids and pandaloids into one monophyletic taxon. Christoffersen (1987) also proposed the new alpheoid families Nauticarididae (to contain *Nauticaris* and *Saron*), Alopidae (to contain *Chorismus, Alope,* and *Caridion*), and Bythocarididae (to contain *Bythocaris, Cryptocheles,* and *Bathyhippolyte*). We have not followed these suggestions, nor have we recognized the families Merhippolytidae and Thoridae recognized by Christoffersen (e.g., Christoffersen 1998).

Christoffersen later (1987) also suggested the recognition of the family Barbouridae (spelling corrected to Barbouriidae by Christoffersen, 1990), to include the genera *Barbouria, Janicea,* and *Parhippolyte.* In his review of caridean shrimps of the *Albatross* Philippine Expedition, Chace (1997), although finding "no clear evidence to support the superfamilial categories suggested by Christoffersen (1987)," found "considerable reason to endorse his [Christoffersen's] establishment of the Barbouriidae." Chace refrained from treating these genera as Barbouriidae in that paper, but we have taken that step here and recognize the Barbouriidae. Inclusion of the family in the superfamily Alpheoida is because of the similarities to hippolytids (all three genera were formerly treated as members of the Hippolytidae).

### Superfamily Crangonoidea

As noted above, Christoffersen (1987) proposed the family Barbouriidae for the genera *Barbouria, Janicea,* and *Parhippolyte* and originally placed the family in the superfamily Crangonoidea. We treat it here as a member of the Alpheoidea because of the similarities to the alpheoid family Hippolytidae (see Chace, 1997:40).

### Superfamily Pandaloidea

Christoffersen (1989) suggested a new classification of this superfamily, wherein he proposed many significant changes. Three new families were proposed (Plesionikidae for the genus *Plesionika,* Heterocarpoididae for the genus *Heterocarpoides,* and Dorodoteidae for the genus *Dorodotes*). In addition, the family Physetocarididae was removed from its own superfamily and placed in the Pandaloidea, and the family Heterocarpidae was recognized. No diagnoses of the new taxa were provided (although character states were given), and we have opted to not recognize these changes for now.

## INFRAORDER ASTACIDEA

Although we are not recognizing the "Macrura Reptantia" as a suborder (see above), for the most part, we have followed the admittedly conservative classification of Holthuis (1991) for the superfamilies and families of the Astacidea (see also Williams, 1988, for classification of commercially important lobster families). Holthuis, who was at the time dealing only with the marine lobsters and so did not include the parastacoids and astacoids, treated marine astacideans as belonging to a single superfamily Nephropoidea containing two families, Thaumastochelidae and Nephropidae. Our classification differs only in the inclusion of the Enoplometopoidea (see below) and Gylpheoidea, the latter placed by Holthuis among the infraorder Palinura (his Palinuridea). Scholtz (1999) recently reviewed the freshwater crayfishes (Astacoidea and Parastacoidea) and argued that they are members of a distinct clade, Astacida, that is not closely related to clawed lobsters. However, strong molecular evidence suggests that clawed lobsters are indeed the sister group to the astacids (Crandall et al., 2000).

### Superfamily Glypheoidea

The primitive family Glypheidae (the only extant family in the Glypheoidea) has been transferred to the Astacidea as per the recommendations of Forest and Saint Laurent (1989). The taxon name, credited to Zittel in Bowman and Abele (1982), has now been credited to the earlier usage by Winckler (M. Hendrickx, pers. comm.), following the usage in Glaessner (1969).

### Superfamily Enoplometopoidea

The genus *Enoplometopus* was assigned its own superfamily and family (Enoplometopoidea, Enoplometopidae) by Saint Laurent (1988).

### Superfamily Nephropoidea

Tshudy and Babcock (1997) examined fossil and extant clawed lobsters and indicated that the family Thaumastochelidae, at least as used previously, may be paraphyletic. We have not taken the extra step of deleting this family (which would result in the former thaumastochelids being treated as Nephropidae), as there was also strong support in their analysis for grouping at least some thaumastochelid genera together (Tshudy and Babcock, 1997: fig. 1).

### Superfamilies Astacoidea and Parastacoidea

Monophyly of the freshwater crayfishes now appears secure based on adult morphology, sperm ultrastructure, embryology, and molecular data (e.g., see Crandall, 1999; Crandall et al., 2000; Scholtz, 1998, 1999). Scholtz (1998, 1999) reviews evolution of the crayfishes and confirms that there are two distinct clades within the group (i.e., within his Astacida) corresponding to the northern hemisphere Astacoidea (families Cambaridae and Astacidae, the latter of which is probably paraphyletic) and the southern hemisphere Parastacoidea (family Parastacidae). Crandall et al. (2000), using over 3000 nucleotides from 3 different genes, have confirmed both the monophyly of the freshwater crayfishes (Astacoidea + Parastacoidea) as well as the monophyly of the astacoid and parastacoid clades. Thus, our current classification is misleading in that these two superfamilies (the Astacoidea and Parastacoidea) are still treated as being of equal rank with three other superfamilies in the Astacidea when in fact they need to be depicted as more closely related to each other than either is to any other astacidean superfamily. Scholtz (1999) also proposes that the crayfishes are not closely related to homarids (not supported by Crandall et al., 2000) but are instead members of "a large group including Thalassinida, Anomala and Brachyura" (see also Scholtz and Richter, 1995). Taylor et al. (1999) added some insights into evolution within the group based on well-preserved fossil material from China.

## INFRAORDER THALASSINIDEA

Monophyly of the thalassinideans is uncertain; at least some morphological and molecular analyses indicate that the group is not monophyletic (e.g., Tudge, 1995; Morrison and Cunningham, 1999). The propensity to construct complex vertical burrows is one character that has been postulated as defining the group (Atkinson and Taylor, 1988; Griffis and Suchanek, 1991; Scholtz and Richter, 1995), as has the presence of a dense row of long setae along the lower margin of the second leg (Poore, 1994, 1997). We have followed the revision by Poore (1994:92) (who also established the family Strahlaxiidae), with the only difference being that some of the authors and dates of some taxa have been changed to earlier usages according to L. Holthuis (pers. comm.). Relationships among the extant superfamilies, families, and genera were suggested by Poore (1994). Poore's resulting classification (1997:92), like ours, does not adequately display all of the relationships suggested by his phylogenetic analysis (Poore, 1994:120). In particular, the Axioidea is the sister group to the Thalassinoidea + Callianassoidea in his phylogeny, whereas in his classification, all three are treated as superfamilies. The family Ctenochelidae is acknowledged by Poore (1994) to be paraphyletic (although Tudge et al., 2000, argued for ctenochelid monophyly). Poore (1997) subsequently addressed three of these families and their relationships in greater detail (Callianideidae Kossman, Micheleidae Sakai, and Thomassiniidae de Saint Laurent).

In the superfamily Callianassoidea, the family Axianassidae was removed by Poore (1994), and the family Ctenochelidae was erected by Manning and Felder (1991). As noted above, Tudge et al. (2000) supported the monophyly of the family Callianassidae and the family Ctenochelidae (while noting that the latter includes, at least in their analysis, the genus *Anacalliax,* considered by some workers to belong to the Callianassidae). In the superfamily Axioidea, Sakai (1992) first established the subfamily Micheleinae, elevated to family level (Micheleidae Sakai) by Poore (1994), and Poore (1994) erected the Strahlaxiidae. Sakai (1999) recently has proposed some rather large-scale revisions within the Callianassidae; his revisions are apparently at odds with other analyses of the same or similar taxa (e.g., see Tudge et al., 2000).

## INFRAORDER PALINURA

Holthuis (1991) referred to this assemblage as the Palinuridea, a spelling that would be consistent with some of our other infraorder names (such as Stenopodidea, Caridea) but not with others (Anomura, Brachyura). We have retained the spelling Palinura. Within the superfamily Palinuroidea, Davie (1990) felt that synaxids were not deserving of separate familial status and synonymized the family Synaxidae with the Palinuridae. However, Holthuis (1991) continued to recognize them as separate families, and we have maintained them as separate families here as well. The family Polychelidae has been recently reviewed and rediagnosed by Galil (2000). Removal of the glypheoids from this infraorder to the Astacidea has been noted above.

## INFRAORDER ANOMURA

Our classification follows McLaughlin's (1983b) fairly closely, with the exception of the use of the family name Pylochelidae replacing Pomatochelidae (following Forest, 1987). McLaughlin (1983a, b) employed the name Anomala De Haan (as had Burkenroad, 1981) rather than Anomura H. Milne Edwards, which had been used by Bowman and Abele (and many other workers). G. Scholtz (pers. comm.) also would prefer this usage (Anomala over Anomura), arguing that when the thalassinoid families are removed the taxon composition changes and thus the name Anomala is the more accurate. Use of Anomala over Anomura was reconsidered and discussed at length by McLaughlin and Holthuis (1985), who pointed out that both names have been used inconsistently in the past and that there are no rules governing the name given to a taxon above the family-group level. Thus, according to McLaughlin and Holthuis, the Rule of Priority need not be applied (Anomala is, strictly speaking, the older of the two names). Furthermore, they argued that, for stability, the name Anomura MacLeay, 1838, should be used for the taxa traditionally considered to belong to this group (lomisoids, galatheoids, paguroids, and hippoids), and we have followed their suggestion. Phylogenetic relationships within the Anomura remain largely unsettled; studies addressing this question include McLaughlin (1983b), Martin and Abele (1986), Cunningham et al. (1992), Tudge (1997b), McLaughlin and Lemaitre (1997, 2000), and Morrison and Cunningham (1999).

McLaughlin (1983a) recognized the unusual nature of *Lomis hirta* and placed it in its own family (Lomidae) and superfamily (Lomoidea) (corrected herein to Lomisidae and Lomisoidea, respectively). McLaughlin (1983b) concluded that the hermit crab families were monophyletic, and she therefore treated all six families as members of the superfamily Paguroidea. This arrangement has been adopted by a variety of workers (e.g., Tudge, 1991; Richter and Scholtz, 1994; Scholtz and Richter, 1995; Tudge, pers. comm.) and seems to us both logical and simple, and we have used it here. In his treatment of the Pylochelidae (treated as Pomatochelidae in Bowman and Abele, 1982), Forest (1987) indicated that the family is more closely allied with the Diogenidae than with other anomuran families, but we have not indicated this alliance pending formal recognition of that relationship.

The family name Lomisidae and the superfamily name Lomisoidea, containing only the monotypic genus *Lomis,* occasionally have been spelled, beginning with Glaessner (1969), as Lomidae and Lomoidea (see especially McLaughlin, 1983a). However, the genus *Lomis* is not a Greek or Latin word, and thus it has no Greek or Latin stem (such as *Lom-*) to which the -idae ending can be added; the original author of *Lomis,* Bouvier, coined the French common name "Lomisinés" for these crabs (G. Poore, pers. comm.). Thus, the preferred spelling for the family is Lomisidae and for the superfamily is Lomisoidea.

A recent analysis of anomuran phylogeny based on mitochondrial DNA gene rearrangements (Morrison and Cunningham, 1999; C. Morrison and C. Cunningham, pers. comm.) largely supports McLaughlin's (1983b) recognition of the major anomuran groups and their phylogeny. According to the findings of Morrison and Cunningham (1999), lithodids are strongly associated with pagurids and together these groups constitute a monophyletic clade (confirming the earlier report by Cunningham et al., 1992). The Hippoidea is also strongly supported as a monophyletic clade, and the Galatheoidea (including both *Aegla* and *Lomis*) is depicted as basal to the remaining Anomura. Thus, a classification based on these data would differ from McLaughlin's (1983a, b) in that the superfamily Lomisoidea would be removed, with the monotypic Lomisidae being placed within the Galatheoidea (which also contains the Aeglidae, Porcellanidae, Galatheidae, and Chirostylidae; see Baba (1988) for a thorough review of the latter family). However, support for this particular node (placement of *Lomis*) was not as strong in the Morrison and Cunningham tree, and indeed C. Morrison (pers. comm.) has suggested that we might be better off depicting a separate lineage for *Aegla* and *Lomis* from the remaining galatheoids. We have for now retained *Lomis* in its own family and superfamily, the Lomisoidea, which we have placed adjacent to the Galatheoidea as a concession to the new data. Similarly, we have moved the Paguroidea closer to the Hippoidea, also reflecting the findings of Morrison and Cunningham (1999). Several workers have discussed the fact that the lithodids (at least some of them) appear to have stemmed from within the Paguridae (Cunningham et al., 1992; Richter and Scholtz, 1994; Tudge, 1991, Tudge et al., 1998; C. Morrison, pers. comm.). Additionally, Cunningham (pers. comm.) suggested a rather close tie between the Aeglidae (restricted to freshwater streams and lakes in temperate South America) and the Lomisidae (a monotypic and exclusively marine family known only from Australia). According to Scholtz and Richter (1995), two groups of the Anomura, hippoids and galatheoids, share the apomorphic character of a telson stretch receptor not found in any other malacostracan group (Scholtz and Richter, 1995, citing Paul, 1989).

In contrast with the phylogenetic hypotheses of McLaughlin (1983b) and Morrison and Cunningham (1999), evidence from sperm ultrastructure (reviewed in Tudge, 1997b) would suggest that the Anomura is not monophyletic, that *Lomis* does not belong to the Anomura *sensu stricta,* that at least some of the thalassinoids are within the Anomura, and that the superfamilies Thalassinoidea, Paguroidea, and Galatheoidea are not monophyletic. Because at this time the bulk of the evidence (i.e., adult morphology combined with molecular sequence and gene arrangement data) seems to support the more conservative approach of McLaughlin (1983b), we have modified our arrangement of anomuran taxa only slightly. Our classification is therefore more in agreement with the findings of Morrison and Cunningham (1999) than with the sperm ultrastructural findings presented by Tudge (1997b).

In the Bowman and Abele (1982) classification, the hermit crab families were divided among two superfamilies, Coenobitoidea and Paguroidea. The Coenobitoidea was removed following the suggestion of McLaughlin (1983b), and the family Coenobitidae is now treated within the superfamily Paguroidea. Thus, our infraorder Anomura contains four superfamilies: Lomisoidea (the distinctness of which is questionable in light of the Morrison and

Cunningham (1999) data, which suggest placement in the galatheoid clade), Galatheoidea, Paguroidea, and Hippoidea (spermatozoal characters of which are described by Tudge et al., 1999). The paguroids (which in our scheme include the former coenobitoids) and hippoids should be considered sister taxa and together are the sister taxon to the Galatheoidea, according to Morrison and Cunningham (1999) and C. Morrison (pers. comm.).

## INFRAORDER BRACHYURA

Subsequent to the Bowman and Abele (1982) classification, there has been relatively widespread use of a scheme first suggested by Guinot (1977, 1978, 1979; see also Saint Laurent, 1979; Guinot and Bouchard, 1998) that recognizes three morphological "grades" of brachyuran crabs (which she called the Podotremata, Heterotremata, and Thoracotremata) based mostly on the coxal vs. sternal location of the male and female genital apertures. Although Abele (1991) and Spears et al. (1992) found no molecular support for these divisions, some spermatological data seemed to support them (e.g., see Jamieson, 1994; Jamieson et al., 1994a, b, 1995). The latter two groups (Heterotremata and Thoracotremata) were treated jointly as the Eubrachyura by Saint Laurent (1980a, b), and various authors (e.g., Schram, 1986) have followed this arrangement as well. At the same time, there is also growing evidence from molecular sequence data (e.g., Spears et al., 1992; Abele and Spears, 1997; Spears and Abele, 1999; Spears, pers. comm.) and from mitochondrial gene rearrangement data (Morrison and Cunningham, 1999; Morrison, pers. comm.) that the true crabs (Brachyura) can be divided into two major clades, one containing the dromiacean families and the other containing all "higher" crabs, and including the raninids. The two ideas are not totally incompatible, but at the same time, they cannot be completely reconciled. The main areas of disagreement concern the limits of the "true" crabs, the placement of several families traditionally thought of as being "primitive" (dromiids and raninids in particular), and the recognition of various assemblages (tribes, sections, etc.) within the major divisions. Evidence brought to bear on these issues has come from many fields, such as larval morphology (e.g., Rice, 1980, 1983, 1988; Martin, 1988, 1991), sperm morphology (e.g., Jamieson, 1991a, b, 1994), adult morphology (e.g., Števčić, 1995, 1998; McLay, 1991, 1999; Guinot and Bouchard, 1998), and molecular sequence data (e.g., Spears et al., 1992).

Guinot (1977, 1978) originally defined the section Podotremata as containing the dromioids, homoloids, raninoids, and tymoloids. The Podotremata was suggested to be monophyletic on the basis of sperm ultrastructure (Jamieson, 1994) and yet paraphyletic on the basis of rRNA sequences (Spears and Abele, 1988; Spears et al., 1992). To quote Guinot and Bouchard (1998), "Monophyly versus paraphyly of the Podotremata and their possible placement as the sister group of the heterotreme-thoracotreme assemblage remain open questions." Within the Podotremata, Guinot (1977, 1978) recognized a subsection Dromiacea to contain two superfamilies, Dromioidea and Homolodromioidea, and a subsection Archaeobrachyura to contain the superfamilies Raninoidea, Homoloidea, and Tymoloidea. The molecular data (e.g., Spears et al., 1992; Spears and Abele, 1999; Morrison and Cunningham, 1999; Spears, pers. comm.) do not support this arrangement. Although one group of crabs, corresponding to the Dromiacea of Guinot and earlier workers, does appear separate from other "higher" crabs, nearly all evidence to date points to the fact that the raninids are not members of this dromioid clade (in contrast with the conclusions of Števčić, 1973, 1995, 1998), and thus the Podotremata cannot be recognized as originally envisioned. Instead, the raninids appear to be basal members of the second "higher crab" clade.

Thus, we have decided to abandon the concept of the Podotremata. The Brachyura is herein depicted as being composed of two major clades. The groups formerly treated as "podotremes" are split, with dromiaceans in one major clade and all other crabs in the other major clade. We are referring to the first clade as the section Dromiacea, a name that has much historical usage and that is well known among brachyuran researchers. This clade (section Dromiacea) is the sister group to all of the higher crab families. In our treatment, it contains the superfamily Homolodromioidea and its sole family Homolodromiidae, the superfamily Dromioidea containing the families Dromiidae and Dynomenidae, and the superfamily Homoloidea containing the Homolidae, Latreilliidae, and Poupiniidae (the latter established by Guinot, 1991).

The second major clade (all other crab families and superfamilies) is then treated collectively as the section Eubrachyura, a name coined by Saint Laurent (1980a, b) for this assemblage (but now including the raninoids, which were excluded by Saint Laurent). We note, however, that Števčić (1973, 1995, 1998) would retain raninids with dromiids, and Jamieson et al. (1994b) argue, based on sperm morphology, against any raninid/higher crab sister group relationship. Inclusion of the raninoids among the Eubrachyura also might be questioned on the basis of the fact that they lack the "sella turcica" of the endophragmal system (see Secretan, 1998). Within this enormous clade Eubrachyura, we are recognizing three subsections. First, we are treating the raninids and their allies (the former tymolids, now treated as the Cyclodorippoidea; see below) as the subsection Raninoida. We could have used for this group the name Archaeobrachyura, a name that has been used previously for the assemblage that contained raninoids, homoloids, and tymoloids (Saint Laurent 1980a, b) while they were still considered members of the "podotreme" lineage. However, use of the name Archaeobrachyura would have been confusing, not only because the constituency and alliances have changed considerably from its original usage by Guinot but because the entire group has been moved to the other major crab clade. We also could have used the older name Gymnopleura, established by Bourne (1922) to accommodate the raninids and still used by some modern workers (e.g., Dai and Yang, 1991). But we have now placed the former tymoloids (now the Cyclodorippoidea) in this subsection with the raninids (which may be a mistake; see below). Hence, our use of the name Raninoida for the subsection. We have credited this higher taxon to the same authority (De Haan) who established the family Raninidae. The other two subsections (the subsections Heterotremata and Thoracotremata), jointly constituting the sister group to the Raninoida, are more or less as envisioned by Guinot (1977, 1978, 1979). Our adoption of Guinot's scheme (minus the Podotremata) has meant that many formerly recognized "tribes" or "sections" among the higher crabs have been removed. This reflects not so much an advance in our knowledge of which families are closely related but rather knowledge concerning which ones are not. For example, the formerly recognized Oxyrhyncha appears to be an artificial assemblage (Števčić and Gore, 1982; Jamieson, 1991a, b, 1994; Spears et al., 1992), and there is no longer any justification for recognizing the Oxystomata, Brachyrhyncha, and other former sections or tribes (e.g., see Guinot, 1977, 1978; Spears et al., 1992; Števčić, 1998). Thus, we have retained several of the crab superfamilies but have removed many of the sections that were found in the Bowman and Abele (1982) classification. Yet acceptance of the sections Heterotremata and Thoracotremata as natural monophyletic lineages is by no means universal. For one thing, Guinot herself never explicitly assigned every known family to one of her sections, leaving some families "orphaned" in her earlier publications. And as noted above, these groups are admittedly (Guinot 1977, 1978) "grades" rather than true monophyletic lineages (or at least, if they are monophyletic, this has yet to be demonstrated, although there are preliminary data from morphology (see below) and from 16S rDNA (Trisha Spears, pers. comm.) that at least the Thoracotremata may have some validity). While usage of these sections has become relatively widespread, it is unfortunate that many families were not explicitly mentioned by Guinot, such that users of her classification have been uncertain as to which families belonged to which section. Schram (1986) provided a more complete list of families (including some known only from fossils).

Concerning monophyly of the Thoracotremata, dissections of the male reproductive tract of a series of freshwater crabs and some marine heterotremes and thoracotremes (during a search for the sister taxon of the freshwater crabs) has indicated that the Thoracotremata is a monophyletic group (Sternberg and Cumberlidge, 2001). One character uniting the thoracotremes is that the distal tracts of the vas deferentia pass through thoracic endosternite 8 and contact the male pleopods via apertures on thoracic sternite 8. The situation in heterotremes is different, with the vas deferens passing through the musculature of endosternite 8 but also through the coxa of pereiopod 5 such that the male sexual tube contacts the pleopods via an aperture on the coxopodite. According to Sternberg et al. (1999), Sternberg and Cumberlidge (2001), and Cumberlidge and Sternberg (pers. comm.), the Eubrachyura (*sensu* Saint Laurent, 1980) are therefore defined by females with sternal vulvae and males with sexual tube outlets that open on the coxa of pereiopod 5. The Thoracotremata constitutes a monophyletic subset of the Eubrachyura characterized by male sexual tube outlets that unambiguously open on the sternum.

Within these last two subsections (Heterotremata and Thoracotremata), many former subfamilies of crabs, notably in the Xanthoidea and Majoidea and some also in the Parthenopoidea, have been elevated to family status based on the publications of several workers (e.g., Serène, 1984, for xanthids; Hendrickx, 1995, for majids). This is an ongoing trend that merely reflects our growing awareness of how incredibly diverse these taxa are.

## SECTION DROMIACEA

In an early version of the updated classification, we had removed the dromiacean crabs from the Brachyura and had placed them instead among the Anomura. Larval characters have suggested this for years (e.g., see Williamson, 1976, 1982; Rice, 1980, 1983; Martin, 1991), so much so that Williamson (1988a, b) invoked an unusual hypothesis of transspecific gene flow to account for it. Molecular (18S rRNA) evidence brought to bear by Spears et al. (1992) seemed to indicate that at least some dromiaceans are indeed closer to the Anomura than to the Brachyura *sensu stricta* based on these preliminary data, and early studies of dromiacean sperm morphology suggested their removal from the Brachyura as well (Jamieson 1990, 1991a). Yet adult morphology has always suggested that dromiids are true crabs (e.g., see Števčić, 1995), and moving the dromiids to the Anomura would raise many additional questions. Should all of the families associated with dromiids (i.e., the former Dromiacea, including dromiids, dynomenids, and homolodromiids) be moved to the Anomura, even though larval and molecular evidence are not in hand for all of them? Is the Dromiacea in fact a valid, monophyletic grouping? If that scheme were accepted, how many other "primitive" families should be moved? The fact that information on larval, molecular, and sperm morphology characters is still lacking for many members of this assemblage, plus more recent molecular data (Spears and Abele, 1999; T. Spears, pers. comm.), eventually led us to keep dromiids with the other "primitive" brachyurans in our section Dromiacea, knowing that by so doing we are continuing to displease students of crab phylogeny who rely mostly on larval characters and that the current arrangement of primitive crabs is not completely in keeping with the molecular evidence in the Spears et al. (1992) study. A detailed discussion of the situation within the Dromiacea can be found in the review of the Dynomenidae by McLay (1999).

### Superfamily Dromioidea

The families Dromiidae and Dynomenidae are still listed as valid families, although based on molecular data (Spears et al., 1992) and sperm morphology (Jamieson, 1994; Jamieson et al., 1995; Guinot et al., 1998), their monophyletic status has been questioned (but see McLay, 1991, 1999; Števčić, 1995). Earlier classifications, some of which have included the Homolidae among the dromiacean families, are reviewed by Števčić (1995), Guinot and Richer de Forges (1995), and McLay (1999). Guinot et al. (1998) argue that the Dromioidea (referred to as Dromiacea in that paper, a lapsus calami, Guinot, pers. comm.), containing the three families Dromiidae, Dynomenidae, and Homolodromiidae, is a valid monophyletic superfamily, although they note the differences separating the homolodromiids. We have maintained the separate status of the homolodromiids (i.e., placing them in their own superfamily Homolodromioidea; see below) in light of the many morphological features of adults that seem to separate them from the dromiids and dynomenids. In doing so, we follow Guinot (1995), even though Guinot and Bouchard (1998) have reverted to treating all three of these families in one superfamily (their Dromiacea). The families were reviewed recently by McLay (1991, Dromiidae; 1999, Dynomenidae) with special regard to their Indo-Pacific members.

### Superfamily Homolodromioidea

Separate superfamily status for the Homolodromiidae appears warranted on the basis of larval and adult morphology (see Martin, 1991; Guinot, 1995). Števčić (1998) considers the homolodromiids the most primitive extant family of brachyuran crabs. The date of Alcock's establishment of the Homolodromiidae has been changed from 1899 to 1900 following the revision by Guinot (1995).

### Superfamily Homoloidea

The alliance of homolids with dromiids has been supported by ultrastructural characters of the sperm (Guinot et al., 1994; see also the extensive review by Guinot and Richer de Forges, 1995). The family Poupiniidae was added by Guinot (1991).

## SECTION EUBRACHYURA, SUBSECTION RANINOIDA

### Superfamily Raninoidea

Within the Raninoidea, the subfamily Symethinae (monogeneric; *Symethis* Goeke) was elevated to family level by Tucker (1998), as had been suggested earlier by Guinot (1993). However, Tucker did not agree with the removal of the subfamily Cyrtorhininae from the Raninidae, which had been suggested as a possibility by Guinot (1993).

### Superfamily Cyclodorippoidea

The superfamily Tymoloidea has been removed and in its place is the superfamily Cyclodorippoidea, as the family name Cyclodorippidae Ortmann has seniority over Tymolidae Alcock, according to Guinot (pers. comm.) and Tavares (1991, 1993). Tavares (1998) also established a new family, the Phyllotymolinidae, within the Cyclodorippoidea. Guinot and Bouchard (1998) continue to recognize the superfamily Cyclodorippoidea (as did Tavares, 1991, 1993, 1998), stating that this was done "for convenience" while at the same time cautioning against possible paraphyly in the assemblage.

Placement of this superfamily with the raninoids in the Raninoida is possibly a mistake; molecular data seem to indicate a placement somewhere between the raninids and the higher eubrachyurans (T. Spears, pers. comm.).

## SECTION EUBRACHYURA, SUBSECTION HETEROTREMATA

### Superfamily Dorippoidea

The family Orithyiidae Dana has been transferred to this superfamily based on the suggestion of Bellwood (1996, 1998; see below).

### Superfamilies Calappoidea and Leucosioidea

The monophyly of the family Calappidae and its constituent subfamilies has been questioned recently. Bellwood (1996, 1998) has recommended that only the families Calappidae and Hepatidae be retained in the superfamily Calappoidea, with the Matutidae joining the leucosiids in the Leucosioidea and with the Orithyiidae transferred to the dorippoids. To some extent, these changes reflect earlier suggestions based on larval (Rice, 1980) and adult (Guinot, 1978; Seridji, 1993) morphology, and there is at least some fossil support for this arrangement as well (Feldmann and Hopkins, 1999; Schweitzer and Feldmann, 2000). Števčić (1983) had earlier suggested recognition of the Matutidae and Orithyidae and their separation from other Calappidae as well. We have followed Bellwood's (1996) recommendations while at the same time not agreeing with her that the Oxystomata be retained. Bellwood's rearrangement of the calappids is not supported by recent molecular data (S. Boyce, unpublished).

## Superfamily Majoidea

Hendrickx (1995, and pers. comm.) brought our attention to the elevation of several majid subfamilies to familial rank, such as the elevation of some inachoid groups by Drach and Guinot (1983), who recognized as families the Inachidae and Inachoididae. We have followed Hendrickx's recognition of former majid subfamilies as families. To some degree, our treatment (and Hendrickx's) of the majoid families follows the seven subfamilies proposed by Griffin and Tranter (1986) in their major revision of the Majidae of the Indo-West Pacific. Additional subfamilies have been proposed by other workers, including Števčić (1994), who disagreed with some of the subdivisions proposed by Griffin and Tranter (1986). Diversity of the former family Majidae is incredibly high, and recognition or treatment of the majoids as a superfamily has been noted or suggested by many earlier workers (e.g., Guinot, 1978; Drach and Guinot, 1983; Števčić, 1994; Clark and Webber, 1991, among others). M. Wicksten (pers. comm.) suggests that, if we elevate some of the former majid subfamilies to the family level, then we should recognize also the family Oregoniidae Garth, 1958, and possibly also the Macrocheiridae Balss, 1929, "for consistency." Indeed, Clark and Webber (1991) proposed recognition of both of these families based on a reevaluation of the larval features of *Macrocheira* and suggested that extant majoids be partitioned among four families: Oregoniidae, Macrocheiridae, Majidae, and Inachidae. Larval morphology indicates the distinct nature, and presumed monophyly, of these groups as well (Pohle and Marques, 2000). We have not taken that step here, feeling that knowledge of larval majoids is still rather incomplete, and we recognize here only the families Epialtidae, Inachidae, Inachoididae, Majidae, Mithracidae, Pisidae, and Tychidae.

Concerning phylogeny among the higher (heterotrematous) crabs, Rice (1983:326) depicts the Majidae (our Majoidea) as basal to the primitive xanthid stock, which in turn gives rise to all other crab families and superfamilies. A recent study based on larval characters (Pohle and Marques, 2000) suggests that, within the Majoidea, the Oregoniinae clade is most basal among those majoid families (or subfamilies) for which larval morphology is known.

## Superfamily Hymenosomatoidea

According to Guinot and Richer de Forges (1997), members of the family Hymenosomatidae (sole member of this superfamily) are thought to be "highly advanced Heterotremata and not Thoracotremata" (Guinot and Richer de Forges, 1997: 454, English abstract). In addition, Guinot and Richer de Forges (1997) revive the idea that the closest relatives of the hymenosomatids may lie among the majoid family Inachoididae. The unusual sperm morphology of one species of the family, as reported by Richer de Forges et al. (1997), would seem to exclude the Hymenosomatidae from the Thoracotremata, and even casts doubts as to the family's inclusion in the Heterotremata.

## Superfamily Parthenopoidea

The superfamily Mimilambroidea and its sole family Mimilambridae, both originally erected by Williams (1979) to contain *Mimilambrus*, have been removed following the suggestion of Ng and Rodriguez (1986) that *Mimilambrus* can be accommodated within the Parthenopidae. Hendrickx (1995) again alerted us to the fact that several former subfamilies of crabs (in this case, former parthenopid subfamilies) had been suggested to be deserving of, or had actually been elevated to, family rank as long ago as 1978 (Guinot, 1978). Although several authors (e.g., Hendrickx, 1999) have attributed the family name Daldorfiidae to M. J. Rathbun, we have found no indication that the taxon was recognized by her. Ng and Rodriguez (1986) recognized the suggested parthenopoid groupings of Guinot as valid families and first used the names Daldorfiidae [as Daldorfidae] and Dairidae, and we have attributed these families to them. We have followed Guinot (1978) and Ng and Rodriguez (1986) and recognize the families Aethridae, Dairidae, Daldorfiidae, and Parthenopidae within a superfamily Parthenopoidea, although Hendrickx (1995) stopped short of treating all of these as valid families.

## Superfamily Retroplumoidea

The family Retroplumidae was given its own superfamily by Saint Laurent (1989), and its placement among the Heterotremata is based on Saint Laurent (1989) and Guinot (pers. comm.)

## Superfamily Cancroidea

The family Cheiragonidae Ortmann, 1893, containing the genera *Telmessus* and *Erimacrus* (formerly treated by most workers as atelecyclids), was resurrected and redescribed by Števčić (1988), and this has been followed by Peter Ng (1998, and pers. comm., 1997; see also Schweitzer and Salva, 2000), and so we have included it here as well.

## Superfamily Portunoidea

The freshwater family Trichodactylidae has now been placed in this superfamily, where it joins the portunids and geryonids, based primarily on a recent morphological analysis (Sternberg et al., 1999; see also below under Potamoidea). Fundamental differences between trichodactylids and other freshwater crabs were recognized by several earlier workers. Rodriguez (1982, 1986, 1992), Magalhães and Türkay (1996a–c), Sternberg (1997), Sternberg et al. (1999), Christoph Schubart (pers. comm.), and Spears et al. (2000) all acknowledge the unique position of the Trichodactylidae and all consider the family monophyletic. The hypothesis that the trichodactylids may represent an independent lineage from any of the other freshwater crab families and that they are descended from portunoid stock is supported by a number of independent studies using morphological data (e.g., Rodriguez, 1982; Magalhães and Türkay, 1996a–c; Sternberg, 1997; Sternberg et al., 1999; and Sternberg and Cumberlidge, in press). Possible corroboration from preliminary molecular evidence (18S, 16S, and 12S rDNA), which is admittedly based on only a handful of freshwater and marine crab species, neither strongly supports nor falsifies this relationship (Abele et al., 1999; Spears et al., 2000). Based on the totality of the evidence available to us, we have transferred the freshwater crab family Trichodactylidae to the marine superfamily Portunoidea.

**Superfamily Bythograeoidea**

Since the discovery of crabs at hydrothermal vents and the erection of a new superfamily and family (Bythograeidae) to accommodate them (Williams, 1980), there has been much discussion concerning the origins and affinities of these crabs (e.g., see Guinot, 1988, 1990; Hessler and Martin, 1989). Williams (1980) noted morphological similarities between bythograeids and portunoids, xanthoids, and potamoids. Guinot (1988) argued for a recent derivation of the hydrothermal crab fauna. Bythograeids are morphologically similar to certain xanthoids, and there are some spermatozoal similarities as well (Tudge et al., 1998). It may be that, at some point, the bythograeids should be transferred to the Xanthoidea. For now, we have left them in their own superfamily.

**Superfamily Xanthoidea**

The former xanthids are now treated as a superfamily containing 11 families, a recognition of the group's diversity that many workers feel is long overdue. The former family Xanthidae contained a wide variety of disparate forms and was the largest single family of the Decapoda, with an estimated 130 genera and over 1,000 species (Rice, 1980; Martin, 1988). Manning and Holthuis (1981) list no fewer than 32 family and subfamily names that have been proposed for various assemblages within the family. Our elevation of the former subfamilies follows mostly the recommendations of Guinot (1977, 1978). A similar subdivision was provided by Serène (1984), although his treatment was restricted to those taxa found in the Red Sea, and so some xanthoid groups (such as the Panopeidae) were not considered by him. Serène (1984) recognized a Xanthoidea containing only five families (Xanthidae, Trapeziidae, Pilumnidae, Carpiliidae, and Menippidae), most with a fairly large number of subfamilies, some of which we are now treating as families. There is recent molecular evidence suggesting that at least some of these former subfamilies are indeed distinct and warrant separate family status (e.g., see Schubart et al., 2000b, for the Panopeidae). Coelho and Coelho Filhol (1993) suggested splitting the former Xanthidae into four families (Carpiliidae, Xanthidae [containing the subfamilies Menippinae, Platyxanthinae, Xanthinae, and Eucratopsinae], Eriphiidae, and Pilumnidae [with subfamilies Trapeziinae and Pilumninae]). One of the problems in elevating the various xanthid groups is that currently there are no published lists of which genera should be included in which family. The field worker who previously could place any xanthoid crab in the Xanthidae is now faced with the rather challenging task of wading through a large amount of primary literature to locate the appropriate family; a further problem is that the primary literature often does not contain all of this information either. Like so many other groups of crustaceans, the "xanthoid" crabs are in need of revision, both taxonomic and phylogenetic (see also Coelho and Coelho Filhol, 1993).

Peter Ng (pers. comm.) feels that the name Eriphiidae MacLeay, 1838, is a senior synonym and should be used instead of Menippidae Ortmann, 1893, for this family, and indeed some workers (e.g., Ng, 1998) have employed the name Eriphiidae. Serène (1984) and other workers have occasionally treated the Eriphiinae as a subfamily of the Menippidae. The family Oziidae Dana, 1852, is apparently a senior synonym of Menippidae as well, as pointed out by Holthuis (1993b), and probably should be used in place of Menippidae if *Ozius* and *Menippe* are both considered members of this group. However, we continue to use Menippidae in this case because the current (fourth) edition of the ICZN allows continued recognition of a name that is enjoying "prevailing use," and in our estimation, replacing Menippidae with Oziidae or Eriphiidae would cause more confusion than maintaining use of Menippidae. Hendrickx (1998) elevated the former goneplacid subfamily Pseudorhombilinae to family status to accommodate six goneplacid-like genera; hence, our inclusion of the family Pseudorhombilidae Alcock, 1900, among the xanthoids.

The Eumedonidae, a family of crabs symbiotic on echinoderms, has at times been recognized as a distinct family (Lim and Ng, 1988; Števčić et al., 1988; and P. Ng, pers. comm.; see Chia and Ng, 2000), and it is often placed within the Xanthoidea, although exactly where it belongs in relation to other crab families is still somewhat uncertain. Most workers are in agreement that early attempts to place it among the parthenopoids were misguided (e.g., see Van Dover et al., 1986; Števčić et al., 1988; Ng and Clark, 1999, 2000) and that it is probably a xanthoid (Števčić et al., 1988). Daniele Guinot (pers. comm.), who earlier listed the family in its own superfamily, the Eumedonoidea Miers (see Guinot, 1985), now also suggests that it might belong in the Xanthoidea, possibly close to the Pilumnidae, a view shared by Van Dover et al. (1986) based on larval evidence. Most recently, Ng and Clark (1999, 2000) have arrived at the conclusion (based primarily on additional strong larval evidence that has accrued since the Van Dover et al. (1986) paper) that eumedonids are simply a subfamily of the Pilumnidae (see also Lim and Ng, 1988). Indeed, Ng (1983) considered it a pilumnid subfamily, as have several other workers (reviewed by Števčić et al., 1988). Yet Chia and Ng (2000) continue to recognize the family. For now, we have continued to treat the Eumedonidae as a separate family with clear affinities to the Pilumnidae, and thus we have placed it with the pilumnids among the xanthoids.

Recognition of *Halimede* as different from other pilumnids goes back at least to the time of Alcock (1898), who recognized the "alliance" Halimedoida. More recent workers (e.g., Serène, 1984:11) have recognized the Halimedinae as a subfamily of the Pilumnidae. Although Bella Galil (pers. comm.) feels that the genus *Halimede* differs sufficiently from other xanthoids to warrant recognition of a separate family, the Halimedidae, we are not aware of any formal treatment or description of the family and how it differs from the other pilumnid groupings. At least some workers (e.g., R. von Sternberg, pers. comm.) would place the Hexapodidae in the Thoracotremata instead of among the xanthoid families in the Heterotremata; von Sternberg also suggests, based primarily on characters of the orbits, that the Goneplacidae may be more closely related to portunids than to other xanthoid families (see also Sternberg and Cumberlidge, in press).

Concerning phylogeny of xanthoid crabs, Rice (1980, 1983) and Martin (1988) have postulated, based on larval features (zoeal and megalopal), that the "Group III" larvae (e.g., *Homalaspis, Ozius, Eriphia*) might be primitive; Martin et al. (1985) suggested that pilumnids might be the least derived assemblage. Guinot (1978) felt that pilumnids and panopeids were more derived than the other groupings. In the current classification, we have simply listed the families alphabetically within the Xanthoidea.

### Superfamily Potamoidea

The higher taxonomy of the freshwater crabs has long been in a state of disarray, and there has been little agreement among authors as to the number of superfamilies and families (e.g., see Cumberlidge, 1999, for a review; Bott, 1970a, b; Pretzmann, 1973; Ng, 1988, 1998; Sternberg et al., 1999; Peter Ng, pers. comm.; Neil Cumberlidge, pers. comm.). Up to 3 superfamilies and 12 families are recognized, depending on the author and also on how far back in the literature one goes. Available higher classifications of the freshwater crabs are based largely on morphological data and, until recently (Rodríguez, 1992; Sternberg, 1997; Sternberg et al., 1999; Sternberg and Cumberlidge, in press), few have been based on cladistic analyses. Many early freshwater crab systematists considered all the world's freshwater crabs to comprise a single monophyletic family, Potamidae. Others (Bott, 1970a, b; Pretzmann, 1973) recognized 11 families and 3 superfamilies, arguing that the group is polyphyletic (or at least paraphyletic) and that similarities represent convergent adaptations of different lineages to similar habitats. Investigations over the past two decades (e.g., Rodríguez, 1982; Ng, 1988; Guinot et al., 1997; Cumberlidge, 1999) have questioned the validity of several families, and these studies continue to reveal the fundamental artificiality of Bott's (1970a,b) 11-family taxonomic arrangement. However, in the absence of a robust phylogenetic study, most authors (including Bowman and Abele, 1982) have adopted their own variant of Bott's classification (albeit reluctantly), and this format is followed here.

Underlying the above taxonomic instability is the unresolved question of the monophyly of the freshwater crabs. A growing body of recent research (Rodríguez, 1992; Sternberg, 1997; Sternberg et al., 1999; Sternberg and Cumberlidge, in press) has falsified the monophyly of the entire group and supports paraphyly with two main lineages. The first lineage includes the Trichodactylidae, which may be descended from some portunoid stock (see above under superfamily Portunoidea), and thus represents an independent line from any of the "potamoid" stock. The second lineage includes the rest of the freshwater crab families. The work of Sternberg et al. (1999), Cumberlidge and Sternberg (1999), Abele et al. (1999), Spears et al. (2000), and Sternberg and Cumberlidge (2000a) indicates that the nontrichodactylid freshwater crabs (all of which are heterotremes) appear to be most closely related to a marine crab clade that includes ocypodids, grapsids, and possibly pinnotherids, with the grapsids providing the best candidate for a sister taxon (an odd result in light of the fact that currently the potamoids are treated as heterotremes whereas the grapsoids are thoracotremes). The hypothesis suggested by Sternberg et al. (1999), that most families of freshwater crabs form a single clade composed of New and Old World lineages, is a departure from the traditional view of the freshwater crab relationships and may lead to further alterations of the higher classification of the group.

Some of the more recent evidence (see especially Abele et al., 1999; Spears et al., 2000) seems to indicate that the freshwater crabs may have arrived via two (and possibly more) invasions. One point of agreement seems to be that the New World pseudothelphusids represent a separate clade from the Old World potamoids. These New World crabs have long been thought to represent an independent lineage (sometimes referred to as the Pseudothelphusoidea; see below) from the rest of the world's freshwater crabs (see also Sternberg and Cumberlidge, 1999). However, even this idea is somewhat controversial concerning whether the trichodactylids belong to the New World clade or represent a separate, independent invasion. Sternberg et al. (1999), citing the works of Magalhães and Türkay

(1996a–c), Rodríguez (1982, 1986, 1992), and Sternberg (1997), feel that there is "strong support for the idea that the Pseudothelphusidae and Trichodactylidae each form a natural group," and Spears and Abele (1999) have suggested that the pseudothelphusids are deserving of superfamily status. Christoph Schubart (pers. comm.) also agrees that the former Potamoidea is polyphyletic, especially as concerns the South American lineages (families Pseudothelphusidae and Trichodactylidae). Our classification is in keeping with most of the above views.

Thus, excluding the trichodactylids, we recognize three superfamilies of freshwater crabs: Potamoidea, Pseudothelphusoidea, and Gercarcinucoidea.

Within the "potamoid" families (superfamily Potamoidea), the families Sinopotamidae and Isolapotamidae have been removed, as both are thought to fall within the limits of the existing Potamidae (Ng, 1988; Dai et al., 1995; Dai, 1997; Dai and Türkay, 1997). Sternberg and Cumberlidge (1999) have recently recognized the monogeneric Platythelphusidae Colossi, 1920, as a distinct potamoid family (see also Cumberlidge et al., 1999; Cumberlidge, 1999) and at the same time suggested that the sister group of the platythelphusids is most likely the East African family Deckeniidae. The Potamonautidae, considered to belong to the Potamidae by Monod (1977, 1980) and Guinot et al. (1997), is recognized as an independent family following the works of Ng (1988), Ng and Takeda (1994), Stewart (1997), Cumberlidge (1999), and Sternberg et al. (1999).

Thus, within the superfamily Potamoidea, we recognize only four families here, all of them Old World groups: Potamidae, Potamonautidae, Deckeniidae, and Platythelphusidae.

### Superfamily Gecarcinucoidea

Only two of the three families originally included in this superfamily by Bott (1970a, b) are recognized here: Gecarcinucidae and Parathelphusidae. The family Sundathelphusidae has been removed, as that family is now considered a junior synonym of the Parathelphusidae (Peter Ng, pers. comm; see also Ng and Sket, 1996; Chia and Ng, 1998). The family Gecarcinucidae, although recognized as being artificial as currently defined and in need of revision (N. Cumberlidge, pers. comm.; and see Cumberlidge, 1987, 1991, 1996a, b, 1999; Cumberlidge and Sachs, 1991), has been retained for now. Membership of the family, as currently defined, is likely to be altered radically in the near future (N. Cumberlidge, pers. comm.). For example, it is possible that the Gecarcinucidae will be shown to be restricted to the Indian subcontinent, Asia, and Australasia (see Cumberlidge, 1999; Martin and Trautwein, in press), and it is not represented on the African continent, despite reports to the contrary (e.g., Bott, 1970a, b). Evidence for maintaining this superfamily (Gecarcinucoidea) and for separating these two families (Gecarcinucidae and Parathelphusidae) from the four families in the Potamoidea is weak and controversial. Nevertheless, we are recognizing the distinctness of the Gecarcinucidae and Parathelphusidae from the four potamoid families until further evidence becomes available.

### Superfamily Pseudothelphusoidea

Originally established by Bott (1970a, b) to include two families, Pseudothelphusidae and Potamocarcinidae, this New World superfamily is now restricted to a single family. The family Potamocarcinidae was removed by Rodríguez (1982), and its species are now included among the Pseudothelphusidae (see Sternberg et al., 1999). The monophyly of the family Pseudothelphusidae appears well established. As noted above, Sternberg et al. (1999), citing the works of Magalhães and Türkay (1996a–c), Rodríguez (1982, 1986, 1992), and Sternberg (1997), feel that there is "strong support for the idea that the Pseudothelphusidae and Trichodactylidae each form a natural group." Spears and Abele (1999) also have suggested that the pseudothelphusids may be deserving of superfamily status, and most workers are in agreement that the pseudothelphusids are a natural (monophyletic) group (T. Spears, pers. comm.; C. Schubart, pers. comm.; Sternberg and Cumberlidge, 1999; Sternberg et al., 1999). We have retained this superfamily and its single family Pseudothelphusidae.

### Superfamily Cryptochiroidea

Finally in the Heterotremata, the correct name for the superfamily and family of the coral gall crabs (Cryptochiroidea and Cryptochiridae, both credited to Paulson) was recognized by Kropp and Manning (1985, 1987), who replaced the name Hapalocarcinidae used previously for this group.

## SECTION EUBRACHYURA, SUBSECTION THORACOTREMATA

### Superfamily Pinnotheroidea

C. Schubart (pers. comm.) believes that the Pinnotheridae "should remain in the Thoracotremata based on evidence from DNA sequencing." Placement of the pinnotherids in the Thoracotremata was also advocated by Števčić (1998) based on morphological features. Thus, the pinnotherids are moved to within the Thoracotremata, although the author of the Thoracotremata does not agree with this placement (Guinot, pers. comm.) and feels that they fit better within the Heterotremata. Within the Pinnotheroidea, it is possible that an additional family will have to be erected to accommodate the genera *Dissodactylus* and *Clypeasterophilus,* which differ morphologically (larval characters) and genetically from other pinnotherids (J. Cuesta, pers. comm.).

## Superfamily Ocypodoidea

Within the Ocypodoidea, Guinot (pers. comm.) questioned the inclusion of the Retroplumidae among the ocypodoids and also among the thoracotremes; she now feels that the family Retroplumidae "probably belongs to the Heterotremata" (where we have now placed it, in its own superfamily following Saint Laurent, 1989). Also within the Ocypodoidea, Guinot (pers. comm.) questions the placement of the Palicidae and suggests that they be listed currently as *incertae sedis*; Guinot and Bouchard (1998) treat them as members of the Heterotremata. C. Schubart (pers. comm.) also questions the placement of the palicids based on results of his 16S mtDNA studies (Schubart et al., 1998). We have left the palicids among the Ocypodoids pending more firm suggestions as to where they might belong. We have also corrected authorship of the family Palicidae to Bouvier from Rathbun (as in Bowman and Abele, 1982, and most other earlier treatments), following the detailed explanation offered in Castro's (2000) revision of the Palicidae of the Indo-West Pacific. The family Camptandriidae Stimpson is recognized by Ng (1988). Schubart (pers. comm.) points out that if we recognize the Camptandriidae, it would be logical also to elevate the other three ocypodid subfamilies (Macropthalminae, Dotillinae, and Heloeciinae) to family level, and apparently there is some preliminary data to support this from zoeal and adult morphology (C. Schubart, pers. comm.). This seems especially logical in light of the finding of Kitaura et al. (1998) that the Camptrandriinae (now Camptrandriidae) is more closely related to the Dotillinae (based on molecular studies) than to any other ocypodid group; however, we have not yet taken that step.

## Superfamily Grapsoidea

It has been suggested that the former grapsid subfamilies (especially the Varuninae) should be elevated to family status based on a combination of morphological, larval, and molecular data (Cuesta and Schubart, 1999; Cuesta et al., 2000; Schubart, 2000a–c; Spivak and Cuesta, 2000; Sternberg and Cumberlidge, 2000b). Schubart, Cuesta, and Felder (in press) review some of these arguments and establish, on the basis of adult and larval morphology and molecular sequence data, the validity of the Glyptograpsidae (containing only *Glyptograpsus* and *Platychirograpsus*); they also review relationships among other former grapsid subfamilies. On the basis of these papers, we recognize as valid families within the Grapsidoidea the Gecarcinidae, Glyptograpsidae, Grapsidae, Plagusiidae, Sesarmidae, and Varunidae. Comparing the families Grapsidae (as restricted; see Schubart, Cuesta, and Felder, in press, and Schubart, Cuesta, and Rodríguez, in press) and Gecarcinidae, Cuesta and Schubart stated (1999: 52) that there is "not a single larval morphological character that consistently distinguishes the Gecarcinidae from the Grapsidae." However, J. Cuesta (pers. comm.) does not feel that the families are closely related and instead feels that larvae of the Gecarcinidae are more similar to larvae of the Varunidae and Sesarmidae.

# CONCLUDING REMARKS

We have thoroughly enjoyed the discussions with, and suggestions from, fellow carcinologists during the compilation and editing of this classification. Doubtless we have pleased and angered some workers more than others in our "final" arrangement. We have been accused of making changes "simply for the sake of change," while at the same time we have been accused of "classificatory paralysis" in our "unwillingness to change." The classification has been criticized as being "nonphylogenetic," while at the same time parts of it have been criticized as relying too heavily on "recent lines of cladistic evidence" (for which read molecular systematics). We accept all such criticisms gladly; they are the signs of a growing and developing field of study and of a field that is of passionate interest to a large number of dedicated workers. We are proud to be your colleagues.

It is our sincere hope that the classification that follows is used primarily as a starting point for future research. By comparing the new classification with that of Bowman and Abele and seeing where changes have, and have not, occurred, and by reading the various dissenting opinions that follow (in Appendix I), we hope that the weaknesses inherent in this classification will be more readily spotted. We further hope that knowledge of these weaknesses will in turn lead to further work on the Crustacea, the planet's most morphologically diverse—and to us, the most interesting—group of organisms.

# CLASSIFICATION OF RECENT CRUSTACEA

Subphylum **Crustacea** Brünnich, 1772
  Class **Branchiopoda** Latreille, 1817
    Subclass **Sarsostraca** Tasch, 1969
      Order **Anostraca** Sars, 1867
        Family **Artemiidae** Grochowski, 1896
          **Branchinectidae** Daday, 1910
          **Branchipodidae** Simon, 1886
          **Chirocephalidae** Daday, 1910
          **Polyartemiidae** Simon, 1886
          **Streptocephalidae** Daday, 1910
          **Thamnocephalidae** Simon, 1886
    Subclass **Phyllopoda** Preuss, 1951
      Order **Notostraca** Sars, 1867
        Family **Triopsidae** Keilhack, 1909
      Order **Diplostraca** Gerstaecker, 1866
        Suborder **Laevicaudata** Linder, 1945
          Family **Lynceidae** Baird, 1845
        Suborder **Spinicaudata** Linder, 1945
          Family **Cyzicidae** Stebbing, 1910
            **Leptestheriidae** Daday, 1923
            **Limnadiidae** Baird, 1849
        Suborder **Cyclestherida** Sars, 1899
          Family **Cyclestheriidae** Sars, 1899
        Suborder **Cladocera** Latreille, 1829
          Infraorder **Ctenopoda** Sars, 1865
            Family **Holopediidae** Sars, 1865
              **Sididae** Baird, 1850
          Infraorder **Anomopoda** Stebbing, 1902
            Family **Bosminidae** Baird, 1845
              **Chydoridae** Stebbing, 1902
              **Daphniidae** Straus, 1820
              **Macrothricidae** Norman & Brady, 1867
          Infraorder **Onychopoda** Sars, 1865
            Family **Cercopagididae** Mordukhai-Boltovskoi, 1968
              **Podonidae** Mordukhai-Boltovskoi, 1968
              **Polyphemidae** Baird, 1845
          Infraorder **Haplopoda** Sars, 1865
            Family **Leptodoridae** Lilljeborg, 1900
  Class **Remipedia** Yager, 1981
      Order **Nectiopoda** Schram, 1986
        Family **Godzilliidae** Schram, Yager & Emerson, 1986
          **Speleonectidae** Yager, 1981
  Class **Cephalocarida** Sanders, 1955
      Order **Brachypoda** Birshteyn, 1960
        Family **Hutchinsoniellidae** Sanders, 1955
  Class **Maxillopoda** Dahl, 1956
    Subclass **Thecostraca** Gruvel, 1905
      Infraclass **Facetotecta** Grygier, 1985
      Infraclass **Ascothoracida** Lacaze-Duthiers, 1880
        Order **Laurida** Grygier, 1987
          Family **Lauridae** Gruvel, 1905
            **Petrarcidae** Gruvel, 1905
            **Synagogidae** Gruvel, 1905
        Order **Dendrogastrida** Grygier, 1987
          Family **Ascothoracidae** Grygier, 1987
            **Ctenosculidae** Thiele, 1925
            **Dendrogastridae** Gruvel, 1905
      Infraclass **Cirripedia** Burmeister, 1834
        Superorder **Acrothoracica** Gruvel, 1905

Order **Pygophora** Berndt, 1907
Family **Cryptophialidae** Gerstaecker, 1866
**Lithoglyptidae** Aurivillius, 1892
Order **Apygophora** Berndt, 1907
Family **Trypetesidae** Stebbing, 1910
Superorder **Rhizocephala** Müller, 1862
Order **Kentrogonida** Delage, 1884
Family **Lernaeodiscidae** Boschma, 1928
**Peltogastridae** Lilljeborg, 1860
**Sacculinidae** Lilljeborg, 1860
Order **Akentrogonida** Häfele, 1911
Family **Chthamalophilidae** Bocquet-Védrine, 1961
**Clistosaccidae** Boschma, 1928
**Duplorbidae** Høeg & Rybakov, 1992
**Mycetomorphidae** Høeg & Rybakov, 1992
**Polysaccidae** Lützen & Takahashi, 1996
**Thompsoniidae** Høeg & Rybakov, 1992
Superorder **Thoracica** Darwin, 1854
Order **Pedunculata** Lamarck, 1818
Suborder **Heteralepadomorpha** Newman, 1987
Family **Anelasmatidae** Gruvel, 1905
**Heteralepadidae** Nilsson-Cantell, 1921
**Koleolepadidae** Hiro, 1933
**Malacolepadidae** Hiro, 1937
**Microlepadidae** Zevina, 1980
**Rhizolepadidae** Zevina, 1980
Suborder **Iblomorpha** Newman, 1987
Family **Iblidae** Leach, 1825
Suborder **Lepadomorpha** Pilsbry, 1916
Family **Lepadidae** Darwin, 1852
**Oxynaspididae** Gruvel, 1905
**Poecilasmatidae** Annandale, 1909
Suborder **Scalpellomorpha** Newman, 1987
Family **Calanticidae** Zevina, 1978
**Lithotryidae** Gruvel, 1905
**Pollicipedidae** Leach, 1817
**Scalpellidae** Pilsbry, 1907
Order **Sessilia** Lamarck, 1818
Suborder **Brachylepadomorpha** Withers, 1923
Family **Neobrachylepadidae** Newman & Yamaguchi, 1995
Suborder **Verrucomorpha** Pilsbry, 1916
Family **Neoverrucidae** Newman, 1989
**Verrucidae** Darwin, 1854
Suborder **Balanomorpha** Pilsbry, 1916
Superfamily **Chionelasmatoidea** Buckeridge, 1983
Family **Chionelasmatidae** Buckeridge, 1983
Superfamily **Pachylasmatoidea** Utinomi, 1968
Family **Pachylasmatidae** Utinomi, 1968
Superfamily **Chthamaloidea** Darwin, 1854
Family **Catophragmidae** Utinomi, 1968
**Chthamalidae** Darwin, 1854
Superfamily **Coronuloidea** Leach, 1817
Family **Chelonibiidae** Pilsbry, 1916
**Coronulidae** Leach, 1817
**Platylepadidae** Newman & Ross, 1976
Superfamily **Tetraclitoidea** Gruvel, 1903
Family **Bathylasmatidae** Newman & Ross, 1971
**Tetraclitidae** Gruvel, 1903
Superfamily **Balanoidea** Leach, 1817
Family **Archaeobalanidae** Newman & Ross, 1976
**Balanidae** Leach, 1817
**Pyrgomatidae** Gray, 1825

Subclass **Tantulocarida** Boxshall & Lincoln, 1983
Family **Basipodellidae** Boxshall & Lincoln, 1983
**Deoterthridae** Boxshall & Lincoln, 1987
**Doryphallophoridae** Huys, 1991
**Microdajidae** Boxshall & Lincoln, 1987
**Onceroxenidae** Huys, 1991
Subclass **Branchiura** Thorell, 1864
Order **Arguloida** Yamaguti, 1963
Family **Argulidae** Leach, 1819
Subclass **Pentastomida** Diesing, 1836
Order **Cephalobaenida** Heymons, 1935
Family **Cephalobaenidae** Fain, 1961
**Reighardiidae** Heymons, 1935
Order **Porocephalida** Heymons, 1935
Family **Armilliferidae** Fain, 1961
**Diesingidae** Fain, 1961
**Linguatulidae** Heymons, 1935
**Porocephalidae** Fain, 1961
**Sambonidae** Fain, 1961
**Sebekiidae** Fain, 1961
**Subtriquetridae** Fain, 1961
Subclass **Mystacocarida** Pennak & Zinn, 1943
Order **Mystacocaridida** Pennak & Zinn, 1943
Family **Derocheilocarididae** Pennak & Zinn, 1943
Subclass **Copepoda** Milne-Edwards, 1840
Infraclass **Progymnoplea** Lang, 1948
Order **Platycopioida** Fosshagen, 1985
Family **Platycopiidae** Sars, 1911
Infraclass **Neocopepoda** Huys & Boxshall, 1991
Superorder **Gymnoplea** Giesbrecht, 1882
Order **Calanoida** Sars, 1903
Family **Acartiidae** Sars, 1900
**Aetideidae** Giesbrecht, 1893
**Arietellidae** Sars, 1902
**Augaptilidae** Sars, 1905
**Bathypontiidae** Brodsky, 1950
**Boholinidae** Fosshagen & Iliffe, 1989
**Calanidae** Dana, 1846
**Candaciidae** Giesbrecht, 1893
**Centropagidae** Giesbrecht, 1893
**Clausocalanidae** Giesbrecht, 1893
**Diaixidae** Sars, 1902
**Diaptomidae** Baird, 1850
**Discoidae** Gordejeva, 1975
**Epacteriscidae** Fosshagen, 1973
**Eucalanidae** Giesbrecht, 1893
**Euchaetidae** Giesbrecht, 1893
**Fosshageniidae** Suárez-Moráles & Iliffe, 1996
**Heterorhabdidae** Sars, 1902
**Hyperbionychidae** Ohtsuka, Roe & Boxshall, 1993
**Lucicutiidae** Sars, 1902
**Mecynoceridae** Andronov, 1973
**Megacalanidae** Sewell, 1947
**Mesaiokeratidae** Matthews, 1961
**Metridinidae** Sars, 1902
**Nullosetigeridae** Soh, Ohtsuka, Imbayashi & Suh, 1999
**Paracalanidae** Giesbrecht, 1893
**Parapontellidae** Giesbrecht, 1893
**Parkiidae** Ferrari & Markhaseva, 1996
**Phaennidae** Sars, 1902
**Pontellidae** Dana, 1852
**Pseudocyclopidae** Giesbrecht, 1893

**Pseudocyclopiidae** Sars, 1902
**Pseudodiaptomidae** Sars, 1902
**Rhincalanidae** Geletin, 1976
**Ridgewayiidae** Wilson, 1958
**Ryocalanidae** Andronov, 1974
**Scolecitrichidae** Giesbrecht, 1893
**Spinocalanidae** Vervoort, 1951
**Stephidae** Sars, 1902
**Sulcanidae** Nicholls, 1945
**Temoridae** Giesbrecht, 1893
**Tharybidae** Sars, 1902
**Tortanidae** Sars, 1902

Superorder **Podoplea** Giesbrecht, 1882
Order **Misophrioida** Gurney, 1933
Family **Misophriidae** Brady, 1878
**Palpophriidae** Boxshall & Jaume, 2000
**Speleophriidae** Boxshall & Jaume, 2000

Order **Cyclopoida** Burmeister, 1834
Family **Archinotodelphyidae** Lang, 1949
**Ascidicolidae** Thorell, 1860
**Buproridae** Thorell, 1859
**Chordeumiidae** Boxshall, 1988
**Cucumaricolidae** Bouligand & Delamare-Deboutteville, 1959
**Cyclopidae** Dana, 1846
**Cyclopinidae** Sars, 1913
**Fratiidae** Ho, Conradi & López-González, 1998
**Lernaeidae** Cobbold, 1879
**Mantridae** Leigh-Sharpe, 1934
**Notodelphyidae** Dana, 1852
**Oithonidae** Dana, 1852
**Ozmanidae** Ho & Thatcher, 1989
**Speleoithonidae** da Rocha & Iliffe, 1991
**Thaumatopsyllidae** Sars, 1913

Order **Gelyelloida** Huys, 1988
Family **Gelyellidae** Rouch & Lescher-Moutoué, 1977

Order **Mormonilloida** Boxshall, 1979
Family **Mormonillidae** Giesbrecht, 1893

Order **Harpacticoida** Sars, 1903
Family **Adenopleurellidae** Huys, 1990
**Aegisthidae** Giesbrecht, 1893
**Ambunguipedidae** Huys, 1990
**Ameiridae** Monard, 1927
**Ancorabolidae** Sars, 1909
**Argestidae** Por, 1986
**Balaenophilidae** Sars, 1910
**Cancrincolidae** Fiers, 1990
**Canthocamptidae** Sars, 1906
**Canuellidae** Lang, 1944
**Cerviniidae** Sars, 1903
**Chappuisiidae** Chappuis, 1940
**Cletodidae** Scott, 1905
**Cletopsyllidae** Huys & Williams, 1989
**Clytemnestridae** Scott, 1909
**Cristacoxidae** Huys, 1990
**Cylindropsyllidae** Sars, 1909
**Darcythompsoniidae** Lang, 1936
**Diosaccidae** Sars, 1906
**Ectinosomatidae** Sars, 1903
**Euterpinidae** Brian, 1921
**Hamondiidae** Huys, 1990
**Harpacticidae** Dana, 1846
**Huntemanniidae** Por, 1986

**Laophontidae** Scott, 1905
**Laophontopsidae** Huys & Willems, 1989
**Latiremidae** Božić, 1969
**Leptastacidae** Lang, 1948
**Leptopontiidae** Lang, 1948
**Longipediidae** Sars, 1903
**Louriniidae** Monard, 1927
**Metidae** Sars, 1910
**Miraciidae** Dana, 1846
**Neobradyidae** Oloffson, 1917
**Normanellidae** Lang, 1944
**Novocriniidae** Huys & Iliffe, 1998
**Orthopsyllidae** Huys, 1990
**Paramesochridae** Lang, 1944
**Parastenheliidae** Lang, 1936
**Parastenocarididae** Chappuis, 1933
**Peltidiidae** Sars, 1904
**Phyllognathopodidae** Gurney, 1932
**Porcellidiidae** Boeck, 1865
**Pseudotachidiidae** Lang, 1936
**Rhizothricidae** Por, 1986
**Rotundiclipeidae** Huys, 1988
**Styracothoracidae** Huys, 1993
**Superornatiremidae** Huys, 1997
**Tachidiidae** Boeck, 1865
**Tegastidae** Sars, 1904
**Tetragonicipitidae** Lang, 1944
**Thalestridae** Sars, 1905
**Thompsonulidae** Lang, 1944
**Tisbidae** Stebbing, 1910

Order **Poecilostomatoida** Thorell, 1859

Family **Anchimolgidae** Humes & Boxshall, 1996
**Anomoclausiidae** Gotto, 1964
**Antheacheridae** Sars, 1870
**Anthessiidae** Humes, 1986
**Bomolochidae** Sumpf, 1871
**Catiniidae** Bocquet & Stock, 1957
**Chitonophilidae** Avdeev & Sirenko, 1991
**Chondracanthidae** Milne Edwards, 1840
**Clausidiidae** Embleton, 1901
**Clausiidae** Giesbrecht, 1895
**Corallovexiidae** Stock, 1975
**Corycaeidae** Dana, 1852
**Echiurophilidae** Delamare-Deboutteville & Nunes-Ruivo, 1955
**Entobiidae** Ho, 1984
**Erebonasteridae** Humes, 1987
**Ergasilidae** von Nordmann, 1832
**Eunicicolidae** Sars, 1918
**Gastrodelphyidae** List, 1889
**Herpyllobiidae** Hansen, 1892
**Intramolgidae** Marchenkov & Boxshall, 1995
**Kelleriidae** Humes & Boxshall, 1996
**Lamippidae** Joliet, 1882
**Lernaeosoleidae** Yamaguti, 1963
**Lichomolgidae** Kossmann, 1877
**Lubbockiidae** Huys & Böttger-Schnack, 1997
**Macrochironidae** Humes & Boxshall, 1996
**Mesoglicolidae** de Zulueta, 1911
**Micrallectidae** Huys, 2001
**Myicolidae** Yamaguti, 1936
**Mytilicolidae** Bocquet & Stock, 1957
**Nereicolidae** Claus, 1875

**Nucellicolidae** Lamb, Boxshall, Mill & Grahame, 1996
**Octopicolidae** Humes & Boxshall, 1996
**Oncaeidae** Giesbrecht, 1893
**Paralubbockiidae** Boxshall & Huys, 1989
**Pharodidae** Illg, 1948
**Philichthyidae** Vogt, 1877
**Philoblennidae** Izawa, 1976
**Phyllodicolidae** Delamare-Deboutteville & Laubier, 1961
**Polyankylidae** Ho & Kim, 1997
**Pseudanthessiidae** Humes & Stock, 1972
**Rhynchomolgidae** Humes & Stock, 1972
**Sabelliphilidae** Gurney, 1927
**Saccopsidae** Lützen, 1964
**Sapphirinidae** Thorell, 1860
**Serpulidicolidae** Stock, 1979
**Shiinoidae** Cressey, 1975
**Spiophanicolidae** Ho, 1984
**Splanchnotrophidae** Norman & Scott, 1906
**Synapticolidae** Humes & Boxshall, 1996
**Synaptiphilidae** Bocquet, 1953
**Taeniacanthidae** Wilson, 1911
**Tegobomolochidae** Avdeev, 1978
**Telsidae** Ho, 1967
**Thamnomolgidae** Humes & Boxshall, 1996
**Tuccidae** Vervoort, 1962
**Urocopiidae** Humes & Stock, 1972
**Vahiniidae** Humes, 1967
**Ventriculinidae** Leigh-Sharpe, 1934
**Xarifiidae** Humes, 1960
**Xenocoelomatidae** Bresciani & Lützen, 1966

Order **Siphonostomatoida** Thorell, 1859

Family **Archidactylinidae** Izawa, 1996
**Artotrogidae** Brady, 1880
**Asterocheridae** Giesbrecht, 1899
**Brychiopontiidae** Humes, 1974
**Caligidae** Burmeister, 1834
**Calverocheridae** Stock, 1968
**Cancerillidae** Giesbrecht, 1897
**Cecropidae** Dana, 1849
**Codobidae** Boxshall & Ohtsuka, 2001
**Coralliomyzontidae** Humes & Stock, 1991
**Dichelesthiidae** Milne Edwards, 1840
**Dichelinidae** Boxshall & Ohtsuka, 2001
**Dinopontiidae** Murnane, 1967
**Dirivultidae** Humes & Dojiri, 1981
**Dissonidae** Yamaguti, 1963
**Ecbathyriontidae** Humes, 1987
**Entomolepididae** Brady, 1899
**Eudactylinidae** Wilson, 1922
**Euryphoridae** Wilson, 1905
**Hatschekiidae** Kabata, 1979
**Hyponeoidae** Heegaard, 1962
**Kroyeriidae** Kabata, 1979
**Lernaeopodidae** Milne Edwards, 1840
**Lernanthropidae** Kabata, 1979
**Megapontiidae** Heptner, 1968
**Micropontiidae** Gooding, 1957
**Nanaspididae** Humes & Cressey, 1959
**Nicothoidae** Dana, 1849
**Pandaridae** Milne Edwards, 1840
**Pennellidae** Burmeister, 1834
**Pontoeciellidae** Giesbrecht, 1895

**Pseudocycnidae** Wilson, 1922
**Rataniidae** Giesbrecht, 1897
**Scottomyzontidae** Ivanenko, Ferrari, & Smurov, 2001
**Sphyriidae** Wilson, 1919
**Sponginticolidae** Topsent, 1928
**Spongiocnizontidae** Stock & Kleeton, 1964
**Stellicomitidae** Humes & Cressey, 1958
**Tanypleuridae** Kabata, 1969
**Trebiidae** Wilson, 1905

Order **Monstrilloida** Sars, 1901
Family **Monstrillidae** Dana, 1849

Class **Ostracoda** Latreille, 1802
Subclass **Myodocopa** Sars, 1866
Order **Myodocopida** Sars, 1866
Suborder **Myodocopina** Sars, 1866
Superfamily **Cypridinoidea** Baird, 1850
Family **Cypridinidae** Baird, 1850
Superfamily **Cylindroleberidoidea** Müller, 1906
Family **Cylindroleberididae** Müller, 1906
Superfamily **Sarsielloidea** Brady & Norman, 1896
Family **Philomedidae** Müller, 1906
**Rutidermatidae** Brady & Norman, 1896
**Sarsiellidae** Brady & Norman, 1896
Order **Halocyprida** Dana, 1853
Suborder **Cladocopina** Sars, 1865
Superfamily **Polycopoidea** Sars, 1865
Family **Polycopidae** Sars, 1865
Suborder **Halocypridina** Dana, 1853
Superfamily **Halocypridoidea** Dana, 1853
Family **Halocyprididae** Dana, 1853
Superfamily **Thaumatocypridoidea** Müller, 1906
Family **Thaumatocyprididae** Müller, 1906
Subclass **Podocopa** Müller, 1894
Order **Platycopida** Sars, 1866
Family **Cytherellidae** Sars, 1866
**Punciidae** Hornibrook, 1949
Order **Podocopida** Sars, 1866
Suborder **Bairdiocopina** Sars, 1865
Superfamily **Bairdioidea** Sars, 1865
Family **Bairdiidae** Sars, 1865
**Bythocyprididae** Maddocks, 1969
Suborder **Cytherocopina** Baird, 1850
Superfamily **Cytheroidea** Baird, 1850
Family **Bythocytheridae** Sars, 1866
**Cytheridae** Baird, 1850
**Cytherideidae** Sars, 1925
**Cytheromatidae** Elofson, 1939
**Cytheruridae** Müller, 1894
**Entocytheridae** Hoff, 1942
**Eucytheridae** Puri, 1954
**Hemicytheridae** Puri, 1953
**Kliellidae** Schäfer, 1945
**Krithidae** Mandelstam, 1958
**Leptocytheridae** Hanai, 1957
**Loxoconchidae** Sars, 1925
**Microcytheridae** Klie, 1938
**Neocytherideidae** Puri, 1957
**Paradoxostomatidae** Brady & Norman, 1889
**Pectocytheridae** Hanai, 1957
**Protocytheridae** Ljubimova, 1956
**Psammocytheridae** Klie, 1938
**Schizocytheridae** Howe, 1961

**Terrestricytheridae** Schornikov, 1969

**Thaerocytheridae** Hazel, 1967

**Trachyleberididae** Sylvester-Bradley, 1948

**Xestoleberididae** Sars, 1928

Suborder **Darwinulocopina** Sohn, 1988

Superfamily **Darwinuloidea** Brady & Norman, 1889

Family **Darwinulidae** Brady & Norman, 1889

Suborder **Cypridocopina** Jones, 1901

Superfamily **Cypridoidea** Baird, 1845

Family **Candonidae** Kaufmann, 1900

**Cyprididae** Baird, 1845

**Ilyocyprididae** Kaufmann, 1900

**Notodromadidae** Kaufmann, 1900

Superfamily **Macrocypridoidea** Müller, 1912

Family **Macrocyprididae** Müller, 1912

Superfamily **Pontocypridoidea** Müller, 1894

Family **Pontocyprididae** Müller, 1894

Suborder **Sigilliocopina** Martens, 1992

Superfamily **Sigillioidea** Mandelstam, 1960

Family **Sigilliidae** Mandelstam, 1960

Class **Malacostraca** Latreille, 1802

Subclass **Phyllocarida** Packard, 1879

Order **Leptostraca** Claus, 1880

Family **Nebaliidae** Samouelle, 1819

**Nebaliopsidae** Hessler, 1984

**Paranebaliidae** Walker-Smith & Poore, 2001

Subclass **Hoplocarida** Calman, 1904

Order **Stomatopoda** Latreille, 1817

Suborder **Unipeltata** Latreille, 1825

Superfamily **Bathysquilloidea** Manning, 1967

Family **Bathysquillidae** Manning, 1967

**Indosquillidae** Manning, 1995

Superfamily **Gonodactyloidea** Giesbrecht, 1910

Family **Alainosquillidae** Moosa, 1991

**Hemisquillidae** Manning, 1980

**Gonodactylidae** Giesbrecht, 1910

**Odontodactylidae** Manning, 1980

**Protosquillidae** Manning, 1980

**Pseudosquillidae** Manning, 1977

**Takuidae** Manning, 1995

Superfamily **Erythrosquilloidea** Manning & Bruce, 1984

Family **Erythrosquillidae** Manning & Bruce, 1984

Superfamily **Lysiosquilloidea** Giesbrecht, 1910

Family **Coronididae** Manning, 1980

**Lysiosquillidae** Giesbrecht, 1910

**Nannosquillidae** Manning, 1980

**Tetrasquillidae** Manning & Camp, 1993

Superfamily **Squilloidea** Latreille, 1802

Family **Squillidae** Latreille, 1802

Superfamily **Eurysquilloidea** Ahyong & Harling, 2000

Family **Eurysquillidae** Manning, 1977

Superfamily **Parasquilloidea** Ahyong & Harling, 2000

Family **Parasquillidae** Manning, 1995

Subclass **Eumalacostraca** Grobben, 1892

Superorder **Syncarida** Packard, 1885

Order **Bathynellacea** Chappuis, 1915

Family **Bathynellidae** Chappuis, 1915

**Parabathynellidae** Noodt, 1965

Order **Anaspidacea** Calman, 1904

Family **Anaspididae** Thomson, 1893

**Koonungidae** Sayce, 1908

**Psammaspididae** Schminke, 1974

**Stygocarididae** Noodt, 1963

Superorder **Peracarida** Calman, 1904

- Order **Spelaeogriphacea** Gordon, 1957
  - Family **Spelaeogriphidae** Gordon, 1957
- Order **Thermosbaenacea** Monod, 1927
  - Family **Halosbaenidae** Monod & Cals, 1988
  - **Monodellidae** Taramelli, 1954
  - **Thermosbaenidae** Monod, 1927
  - **Tulumellidae** Wagner, 1994
- Order **Lophogastrida** Sars, 1870
  - Family **Eucopiidae** Sars, 1885
  - **Lophogastridae** Sars, 1870
- Order **Mysida** Haworth, 1825
  - Family **Lepidomysidae** Clarke, 1961
  - **Mysidae** Haworth, 1825
  - **Petalophthalmidae** Czerniavsky, 1882
  - **Stygiomysidae** Caroli, 1937
- Order **Mictacea** Bowman, Garner, Hessler, Iliffe & Sanders, 1985
  - Family **Hirsutiidae** Sanders, Hessler & Garner, 1985
  - **Mictocarididae** Bowman & Iliffe, 1985
- Order **Amphipoda** Latreille, 1816
  - Suborder **Gammaridea** Latreille, 1802
    - Family **Acanthogammaridae** Garjajeff, 1901
    - **Acanthonotozomellidae** Coleman & Barnard, 1991
    - **Allocrangonyctidae** Holsinger, 1989
    - **Amathillopsidae** Pirlot, 1934
    - **Ampeliscidae** Costa, 1857
    - **Amphilochidae** Boeck, 1871
    - **Ampithoidae** Stebbing, 1899
    - **Anamixidae** Stebbing, 1897
    - **Anisogammaridae** Bousfield, 1977
    - **Aoridae** Walker, 1908
    - **Argissidae** Walker, 1904
    - **Aristiidae** Lowry & Stoddart, 1997
    - **Artesiidae** Holsinger, 1980
    - **Bateidae** Stebbing, 1906
    - **Biancolinidae** Barnard, 1972
    - **Bogidiellidae** Hertzog, 1936
    - **Bolttsiidae** Barnard & Karaman, 1987
    - **Calliopidae** Sars, 1893
    - **Carangoliopsidae** Bousfield, 1977
    - **Cardenioidae** Barnard & Karaman, 1987
    - **Caspicolidae** Birstein, 1945
    - **Ceinidae** Barnard, 1972
    - **Cheidae** Thurston, 1982
    - **Cheluridae** Allman, 1847
    - **Clarenciidae** Barnard & Karaman, 1987
    - **Colomastigidae** Stebbing, 1899
    - **Condukiidae** Barnard & Drummond, 1982
    - **Corophiidae** Leach, 1814
    - **Crangonyctidae** Bousfield, 1973
    - **Cressidae** Stebbing, 1899
    - **Cyphocarididae** Lowry & Stoddart, 1997
    - **Cyproideidae** Barnard, 1974
    - **Dexaminidae** Leach, 1814
    - **Didymocheliidae** Bellan-Santini & Ledoyer, 1986
    - **Dikwidae** Coleman & Barnard, 1991
    - **Dogielinotidae** Gurjanova, 1953
    - **Dulichiidae** Dana, 1849
    - **Endevouridae** Lowry & Stoddart, 1997
    - **Eophliantidae** Sheard, 1936
    - **Epimeriidae** Boeck, 1871

**Eusiridae** Stebbing, 1888
**Exoedicerotidae** Barnard & Drummond, 1982
**Gammaracanthidae** Bousfield, 1989
**Gammarellidae** Bousfield, 1977
**Gammaridae** Latreille, 1802
**Gammaroporeiidae** Bousfield, 1979
**Hadziidae** Karaman, 1943
**Haustoriidae** Stebbing, 1906
**Hyalellidae** Bulycheva, 1957
**Hyalidae** Bulycheva, 1957
**Hyperiopsidae** Bovallius, 1886
**Iciliidae** Dana, 1849
**Ipanemidae** Barnard & Thomas, 1988
**Iphimediidae** Boeck, 1871
**Isaeidae** Dana, 1853
**Ischyroceridae** Stebbing, 1899
**Kuriidae** Walker & Scott, 1903
**Laphystiidae** Sars, 1893
**Laphystiopsidae** Stebbing, 1899
**Lepechinellidae** Schellenberg, 1926
**Leucothoidae** Dana, 1852
**Liljeborgiidae** Stebbing, 1899
**Lysianassidae** Dana, 1849
**Macrohectopidae** Sowinsky, 1915
**Maxillipiidae** Ledoyer, 1973
**Megaluropidae** Thomas & Barnard, 1986
**Melitidae** Bousfield, 1973
**Melphidippidae** Stebbing, 1899
**Mesogammaridae** Bousfield, 1977
**Metacrangonyctidae** Boutin & Missouli, 1988
**Micruropidae** Kamaltynov, 1999
**Najnidae** Barnard, 1972
**Neomegamphopidae** Myers, 1981
**Neoniphargidae** Bousfield, 1977
**Nihotungidae** Barnard, 1972
**Niphargidae** Bousfield, 1977
**Ochlesidae** Stebbing, 1910
**Odiidae** Coleman & Barnard, 1991
**Oedicerotidae** Lilljeborg, 1865
**Opisidae** Lowry & Stoddart, 1995
**Pachyschesidae** Kamaltynov, 1999
**Pagetinidae** Barnard, 1931
**Paracalliopidae** Barnard & Karaman, 1982
**Paracrangonyctidae** Bousfield, 1982
**Paraleptamphopidae** Bousfield, 1983
**Paramelitidae** Bousfield, 1977
**Pardaliscidae** Boeck, 1871
**Perthiidae** Williams & Barnard, 1988
**Phliantidae** Stebbing, 1899
**Phoxocephalidae** Sars, 1891
**Phoxocephalopsidae** Barnard & Drummond, 1982
**Phreatogammaridae** Bousfield, 1982
**Platyischnopidae** Barnard & Drummond, 1979
**Pleustidae** Buchholz, 1874
**Plioplateidae** Barnard, 1978
**Podoceridae** Leach, 1814
**Podoprionidae** Lowry & Stoddart, 1996
**Pontogammaridae** Bousfield, 1977
**Pontoporeiidae** Dana, 1853
**Priscomilitaridae** Hirayama, 1988
**Pseudamphilochidae** Schellenberg, 1931
**Pseudocrangonyctidae** Holsinger, 1989

**Salentinellidae** Bousfield, 1977
**Scopelocheiridae** Lowry & Stoddart, 1997
**Sebidae** Walker, 1908
**Sinurothoidae** Ren, 1999
**Stegocephalidae** Dana, 1853
**Stenothoidae** Boeck, 1871
**Sternophysingidae** Holsinger, 1992
**Stilipedidae** Holmes, 1908
**Synopiidae** Dana, 1853
**Talitridae** Rafinesque, 1815
**Temnophliantidae** Griffiths, 1975
**Trischizostomatidae** Lilljeborg, 1865
**Tulearidae** Ledoyer, 1979
**Typhlogammaridae,** Bousfield, 1977
**Uristidae** Hurley, 1963
**Urohaustoriidae** Barnard & Drummond, 1982
**Urothoidae** Bousfield, 1978
**Valettidae** Stebbing, 1888
**Vicmusiidae** Just, 1990
**Vitjazianidae** Birstein & Vinogradov, 1955
**Wandinidae** Lowry & Stoddart, 1990
**Zobrachoidae** Barnard & Drummond, 1982

Suborder **Caprellidea** Leach, 1814
Infraorder **Caprellida** Leach, 1814
Superfamily **Caprelloidea** Leach, 1814
Family **Caprellidae** Leach, 1814
**Caprellinoididae** Laubitz, 1993
**Caprogammaridae** Kudrjaschov & Vassilenko, 1966
**Paracercopidae** Vassilenko, 1968
**Pariambidae** Laubitz, 1993
**Protellidae** McCain, 1970
Superfamily **Phtisicoidea** Vassilenko, 1968
Family **Phtisicidae** Vassilenko, 1968
Infraorder **Cyamida** Rafinesque, 1815
Family **Cyamidae** Rafinesque, 1815

Suborder **Hyperiidea** Milne Edwards, 1830
Infraorder **Physosomata** Pirlot, 1929
Superfamily **Scinoidea** Stebbing, 1888
Family **Archaeoscinidae** Stebbing, 1904
**Mimonectidae** Bovallius, 1885
**Proscinidae** Pirlot, 1933
**Scinidae** Stebbing, 1888
Superfamily **Lanceoloidea** Bovallius, 1887
Family **Chuneolidae** Woltereck, 1909
**Lanceolidae** Bovallius, 1887
**Microphasmatidae** Stephensen & Pirlot, 1931
Infraorder **Physocephalata** Bowman & Gruner, 1973
Superfamily **Vibilioidea** Dana, 1853
Family **Cystisomatidae** Willemoes-Suhm, 1875
**Paraphronimidae** Bovallius, 1887
**Vibiliidae** Dana, 1853
Superfamily **Phronimoidea** Rafinesque, 1815
Family **Dairellidae** Bovallius, 1887
**Hyperiidae** Dana, 1853
**Phronimidae** Rafinesque, 1815
**Phrosinidae** Dana, 1853
Superfamily **Lycaeopsoidea** Chevreux, 1913
Family **Lycaeopsidae** Chevreux, 1913
Superfamily **Platysceloidea** Bate, 1862
Family **Anapronoidae** Bowman & Gruner, 1973
**Lycaeidae** Claus, 1879
**Oxycephalidae** Dana, 1853

**Parascelidae** Bate, 1862
**Platyscelidae** Bate, 1862
**Pronoidae** Dana, 1853
Suborder **Ingolfiellidea** Hansen, 1903
Family **Ingolfiellidae** Hansen, 1903
**Metaingolfiellidae** Ruffo, 1969
Order **Isopoda** Latreille, 1817
Suborder **Phreatoicidea** Stebbing, 1893
Family **Amphisopodidae** Nicholls, 1943
**Nichollsiidae** Tiwari, 1955
**Phreatoicidae** Chilton, 1891
Suborder **Anthuridea** Monod, 1922
Family **Antheluridae** Poore & Lew Ton, 1988
**Anthuridae** Leach, 1814
**Expanathuridae** Poore, 2001
**Hyssuridae** Wägele, 1981
**Leptanthuridae** Poore, 2001
**Paranthuridae** Menzies & Glynn, 1968
Suborder **Microcerberidea** Lang, 1961
Family **Atlantasellidae** Sket, 1980
**Microcerberidae** Karaman, 1933
Suborder **Flabellifera** Sars, 1882
Family **Aegidae** White, 1850
**Ancinidae** Dana, 1852
**Anuropidae** Stebbing, 1893
**Bathynataliidae** Kensley, 1978
**Cirolanidae** Dana, 1852
**Corallanidae** Hansen, 1890
**Cymothoidae** Leach, 1814
**Gnathiidae** Leach, 1814
**Hadromastacidae** Bruce & Müller, 1991
**Keuphyliidae** Bruce, 1980
**Limnoriidae** White, 1850
**Phoratopodidae** Hale, 1925
**Plakarthriidae** Hansen, 1905
**Protognathiidae** Wägele & Brandt, 1988
**Serolidae** Dana, 1852
**Sphaeromatidae** Latreille, 1825
**Tecticepitidae** Iverson, 1982
**Tridentellidae** Bruce, 1984
Suborder **Asellota** Latreille, 1802
Superfamily **Aselloidea** Latreille, 1802
Family **Asellidae** Latreille, 1802
**Stenasellidae** Dudich, 1924
Superfamily **Stenetrioidea** Hansen, 1905
Family **Pseudojaniridae** Wilson, 1986
**Stenetriidae** Hansen, 1905
Superfamily **Janiroidea** Sars, 1897
Family **Acanthaspidiidae** Menzies, 1962
**Dendrotiidae** Vanhöffen, 1914
**Desmosomatidae** Sars, 1899
**Echinothambematidae** Menzies, 1956
**Haplomunnidae** Wilson, 1976
**Haploniscidae** Hansen, 1916
**Ischnomesidae** Hansen, 1916
**Janirellidae** Menzies, 1956
**Janiridae** Sars, 1897
**Joeropsididae** Nordenstam, 1933
**Katianiridae** Svavarsson, 1987
**Macrostylidae** Hansen, 1916
**Mesosignidae** Schultz, 1969
**Microparasellidae** Karaman, 1933

**Mictosomatidae** Wolff, 1965
**Munnidae** Sars, 1897
**Munnopsididae** Sars, 1869
**Nannoniscidae** Hansen, 1916
**Paramunnidae** Vanhöffen, 1914
**Pleurocopidae** Fresi & Schiecke, 1972
**Santiidae** Wilson, 1987
**Thambematidae** Stebbing, 1913
Superfamily **Gnathostenetroidoidea** Kussakin, 1967
Family **Gnathostenetroididae** Kussakin, 1967
**Protojaniridae** Fresi, Idato & Scipione, 1980
**Vermectiadidae** Just & Poore, 1992
Suborder **Calabozoida** Van Lieshout, 1983
Family **Calabozoidae** Van Lieshout, 1983
Suborder **Valvifera** Sars, 1882
Family **Antarcturidae** Poore, 2001
**Arcturidae** Dana, 1849
**Arcturididae** Poore, 2001
**Austrarcturellidae** Poore & Bardsley, 1992
**Chaetiliidae** Dana, 1849
**Holidoteidae** Wägele, 1989
**Holognathidae** Thomson, 1904
**Idoteidae** Samouelle, 1819
**Pseudidotheidae** Ohlin, 1901
**Rectarcturidae** Poore, 2001
**Xenarcturidae** Sheppard, 1957
Suborder **Epicaridea** Latreille, 1831
Superfamily **Bopyroidea** Rafinesque, 1815
Family **Bopyridae** Rafinesque, 1815
**Dajidae** Giard & Bonnier, 1887
**Entoniscidae** Kossmann, 1881
Superfamily **Cryptoniscoidea** Kossmann, 1880
Family **Asconiscidae** Bonnier, 1900
**Cabiropidae** Giard & Bonnier, 1887
**Crinoniscidae** Bonnier, 1900
**Cryptoniscidae** Kossmann, 1880
**Cyproniscidae** Bonnier, 1900
**Fabidae** Danforth, 1963
**Hemioniscidae** Bonnier, 1900
**Podasconidae** Bonnier, 1900
Suborder **Oniscidea** Latreille, 1802
Family **Dubioniscidae** Schultz, 1995
**Helelidae** Ferrara, 1977
**Irmaosidae** Ferrara & Taiti, 1983
**Pseudarmadillidae** Vandel, 1973
**Scleropactidae** Verhoeff, 1938
Infraorder **Tylomorpha** Vandel, 1943
Family **Tylidae** Dana, 1852
Infraorder **Ligiamorpha** Vandel, 1943
Section **Diplocheta** Vandel, 1957
Family **Ligiidae** Leach, 1814
**Mesoniscidae** Verhoeff, 1908
Section **Synocheta** Legrand, 1946
Superfamily **Trichoniscoidea** Sars, 1899
Family **Buddelundiellidae** Verhoeff, 1930
**Trichoniscidae** Sars, 1899
Superfamily **Styloniscoidea** Vandel, 1952
Family **Schoebliidae** Verhoeff, 1938
**Styloniscidae** Vandel, 1952
**Titaniidae** Verhoeff, 1938
**Tunanoniscidae** Borutskii, 1969
Section **Crinocheta** Legrand, 1946

Superfamily **Oniscoidea** Latreille, 1802
Family **Bathytropidae** Vandel, 1952
**Berytoniscidae** Vandel, 1973
**Detonidae** Budde-Lund, 1906
**Halophilosciidae** Verhoeff, 1908
**Olibrinidae** Vandel, 1973
**Oniscidae** Latreille, 1802
**Philosciidae** Kinahan, 1857
**Platyarthridae** Vandel, 1946
**Pudeoniscidae** Lemos de Castro, 1973
**Rhyscotidae** Budde-Lund, 1908
**Scyphacidae** Dana, 1852
**Speleoniscidae** Vandel, 1948
**Sphaeroniscidae** Vandel, 1964
**Stenoniscidae** Budde-Lund, 1904
**Tendosphaeridae** Verhoeff, 1930
Superfamily **Armadilloidea** Brandt, 1831
Family **Actaeciidae** Vandel, 1952
**Armadillidae** Brandt, 1831
**Armadillidiidae** Brandt, 1833
**Atlantidiidae** Arcangeli, 1954
**Balloniscidae** Vandel, 1963
**Cylisticidae** Verhoeff, 1949
**Eubelidae** Budde-Lund, 1904
**Periscyphicidae** Ferrara, 1973
**Porcellionidae** Brandt, 1831
**Trachelipodidae** Strouhal, 1953
Order **Tanaidacea** Dana, 1849
Suborder **Tanaidomorpha** Sieg, 1980
Superfamily **Tanaoidea** Dana, 1849
Family **Tanaidae** Dana, 1849
Superfamily **Paratanaoidea** Lang, 1949
Family **Anarthruridae** Lang, 1971
**Leptochelidae** Lang, 1973
**Nototanaidae** Sieg, 1976
**Paratanaidae** Lang, 1949
**Pseudotanaidae** Sieg, 1976
**Pseudozeuxidae** Sieg, 1982
**Typhlotanaidae** Sieg, 1986
Suborder **Neotanaidomorpha** Sieg, 1980
Family **Neotanaidae** Lang, 1956
Suborder **Apseudomorpha** Sieg, 1980
Superfamily **Apseudoidea** Leach, 1814
Family **Anuropodidae** Băcescu, 1980
**Apseudellidae** Gutu, 1972
**Apseudidae** Leach, 1814
**Gigantapseudidae** Kudinova-Pasternak, 1978
**Kalliapseudidae** Lang, 1956
**Metapseudidae** Lang, 1970
**Pagurapseudidae** Lang, 1970
**Parapseudidae** Gutu, 1981
**Sphyrapidae** Gutu, 1980
**Tanapseudidae** Băcescu, 1978
**Tanzanapseudidae** Băcescu, 1975
**Whiteleggiidae** Gutu, 1972
Order **Cumacea** Krøyer, 1846
Family **Bodotriidae** Scott, 1901
**Ceratocumatidae** Calman, 1905
**Diastylidae** Bate, 1856
**Gynodiastylidae** Stebbing, 1912
**Lampropidae** Sars, 1878
**Leuconidae** Sars, 1878

**Nannastacidae** Bate, 1866
**Pseudocumatidae** Sars, 1878

Superorder **Eucarida** Calman, 1904
Order **Euphausiacea** Dana, 1852
Family **Bentheuphausiidae** Colosi, 1917
**Euphausiidae** Dana, 1852
Order **Amphionidacea** Williamson, 1973
Family **Amphionididae** Holthuis, 1955
Order **Decapoda** Latreille, 1802
Suborder **Dendrobranchiata** Bate, 1888
Superfamily **Penaeoidea** Rafinesque, 1815
Family **Aristeidae** Wood-Mason, 1891
**Benthesicymidae** Wood-Mason, 1891
**Penaeidae** Rafinesque, 1815
**Sicyoniidae** Ortmann, 1898
**Solenoceridae** Wood-Mason, 1891
Superfamily **Sergestoidea** Dana, 1852
Family **Luciferidae** de Haan, 1849
**Sergestidae** Dana, 1852
Suborder **Pleocyemata** Burkenroad, 1963
Infraorder **Stenopodidea** Claus, 1872
Family **Spongicolidae** Schram, 1986
**Stenopodidae** Claus, 1872
Infraorder **Caridea** Dana, 1852
Superfamily **Procaridoidea** Chace & Manning, 1972
Family **Procarididae** Chace & Manning, 1972
Superfamily **Galatheacaridoidea** Vereshchaka, 1997
Family **Galatheacarididae** Vereshchaka, 1997
Superfamily **Pasiphaeoidea** Dana, 1852
Family **Pasiphaeidae** Dana, 1852
Superfamily **Oplophoroidea** Dana, 1852
Family **Oplophoridae** Dana, 1852
Superfamily **Atyoidea** de Haan, 1849
Family **Atyidae** de Haan, 1849
Superfamily **Bresilioidea** Calman, 1896
Family **Agostocarididae** Hart & Manning, 1986
**Alvinocarididae** Christoffersen, 1986
**Bresiliidae** Calman, 1896
**Disciadidae** Rathbun, 1902
**Mirocarididae** Vereshchaka, 1997
Superfamily **Nematocarcinoidea** Smith, 1884
Family **Eugonatonotidae** Chace, 1937
**Nematocarcinidae** Smith, 1884
**Rhynchocinetidae** Ortmann, 1890
**Xiphocarididae** Ortmann, 1895
Superfamily **Psalidopodoidea** Wood-Mason & Alcock, 1892
Family **Psalidopodidae** Wood-Mason & Alcock, 1892
Superfamily **Stylodactyloidea** Bate, 1888
Family **Stylodactylidae** Bate, 1888
Superfamily **Campylonotoidea** Sollaud, 1913
Family **Bathypalaemonellidae** de Saint Laurent, 1985
**Campylonotidae** Sollaud, 1913
Superfamily **Palaemonoidea** Rafinesque, 1815
Family **Anchistioididae** Borradaile, 1915
**Desmocarididae** Borradaile, 1915
**Euryrhynchidae** Holthuis, 1950
**Gnathophyllidae** Dana, 1852
**Hymenoceridae** Ortmann, 1890
**Kakaducarididae** Bruce, 1993
**Palaemonidae** Rafinesque, 1815
**Typhlocarididae** Annandale & Kemp, 1913
Superfamily **Alpheoidea** Rafinesque, 1815

Family **Alpheidae** Rafinesque, 1815
**Barbouriidae** Christoffersen, 1987
**Hippolytidae** Dana, 1852
**Ogyrididae** Holthuis, 1955
Superfamily **Processoidea** Ortmann, 1890
Family **Processidae** Ortmann, 1890
Superfamily **Pandaloidea** Haworth, 1825
Family **Pandalidae** Haworth, 1825
**Thalassocarididae** Bate, 1888
Superfamily **Physetocaridoidea** Chace, 1940
Family **Physetocarididae** Chace, 1940
Superfamily **Crangonoidea** Haworth, 1825
Family **Crangonidae** Haworth, 1825
**Glyphocrangonidae** Smith, 1884
Infraorder **Astacidea** Latreille, 1802
Superfamily **Glypheoidea** Winkler, 1883
Family **Glypheidae** Winkler, 1883
Superfamily **Enoplometopoidea** de Saint Laurent, 1988
Family **Enoplometopidae** de Saint Laurent, 1988
Superfamily **Nephropoidea** Dana, 1852
Family **Nephropidae** Dana, 1852
**Thaumastochelidae** Bate, 1888
Superfamily **Astacoidea** Latreille, 1802
Family **Astacidae** Latreille, 1802
**Cambaridae** Hobbs, 1942
Superfamily **Parastacoidea** Huxley, 1879
Family **Parastacidae** Huxley, 1879
Infraorder **Thalassinidea** Latreille, 1831
Superfamily **Thalassinoidea** Latreille, 1831
Family **Thalassinidae** Latreille, 1831
Superfamily **Callianassoidea** Dana, 1852
Family **Callianassidae** Dana, 1852
**Callianideidae** Kossmann, 1880
**Ctenochelidae** Manning & Felder, 1991
**Laomediidae** Borradaile, 1903
**Thomassiniidae** de Saint Laurent, 1979
**Upogebiidae** Borradaile, 1903
Superfamily **Axioidea** Huxley, 1879
Family **Axiidae** Huxley, 1879
**Calocarididae** Ortmann, 1891
**Micheleidae** Sakai, 1992
**Strahlaxiidae** Poore, 1994
Infraorder **Palinura** Latreille, 1802
Superfamily **Eryonoidea** de Haan, 1841
Family **Polychelidae** Wood-Mason, 1874
Superfamily **Palinuroidea** Latreille, 1802
Family **Palinuridae** Latreille, 1802
**Scyllaridae** Latreille, 1825
**Synaxidae** Bate, 1881
Infraorder **Anomura** MacLeay, 1838
Superfamily **Lomisoidea** Bouvier, 1895
Family **Lomisidae** Bouvier, 1895
Superfamily **Galatheoidea** Samouelle, 1819
Family **Aeglidae** Dana, 1852
**Chirostylidae** Ortmann, 1892
**Galatheidae** Samouelle, 1819
**Porcellanidae** Haworth, 1825
Superfamily **Hippoidea** Latreille, 1825
Family **Albuneidae** Stimpson, 1858
**Hippidae** Latreille, 1825
Superfamily **Paguroidea** Latreille, 1802
Family **Coenobitidae** Dana, 1851

**Diogenidae** Ortmann, 1892
**Lithodidae** Samouelle, 1819
**Paguridae** Latreille, 1802
**Parapaguridae** Smith, 1882
**Pylochelidae** Bate, 1888

Infraorder **Brachyura** Latreille, 1802
Section **Dromiacea** de Haan, 1833
Superfamily **Homolodromioidea** Alcock, 1900
Family **Homolodromiidae** Alcock, 1900
Superfamily **Dromioidea** de Haan, 1833
Family **Dromiidae** de Haan, 1833
**Dynomenidae** Ortmann, 1892
Superfamily **Homoloidea** de Haan, 1839
Family **Homolidae** de Haan, 1839
**Latreilliidae** Stimpson, 1858
**Poupiniidae** Guinot, 1991
Section **Eubrachyura** de Saint Laurent, 1980
Subsection **Raninoida** de Haan, 1839
Superfamily **Raninoidea** de Haan, 1839
Family **Raninidae** de Haan, 1839
**Symethidae** Goeke, 1981
Superfamily **Cyclodorippoidea** Ortmann, 1892
Family **Cyclodorippidae** Ortmann, 1892
**Cymonomidae** Bouvier, 1897
**Phyllotymolinidae** Tavares, 1998
Subsection **Heterotremata** Guinot, 1977
Superfamily **Dorippoidea** MacLeay, 1838
Family **Dorippidae** MacLeay, 1838
**Orithyiidae** Dana, 1853
Superfamily **Calappoidea** Milne Edwards, 1837
Family **Calappidae** Milne Edwards, 1837
**Hepatidae** Stimpson, 1871
Superfamily **Leucosioidea** Samouelle, 1819
Family **Leucosiidae** Samouelle, 1819
**Matutidae** de Hann, 1841
Superfamily **Majoidea** Samouelle, 1819
Family **Epialtidae** MacLeay, 1838
**Inachidae** MacLeay, 1838
**Inachoididae** Dana, 1851
**Majidae** Samouelle, 1819
**Mithracidae** Balss, 1929
**Pisidae** Dana, 1851
**Tychidae** Dana, 1851
Superfamily **Hymenosomatoidea** MacLeay, 1838
Family **Hymenosomatidae** MacLeay, 1838
Superfamily **Parthenopoidea** MacLeay, 1838
Family **Aethridae** Dana, 1851
**Dairidae** Ng & Rodriguez, 1986
**Daldorfiidae** Ng & Rodriguez, 1986
**Parthenopidae** MacLeay, 1838
Superfamily **Retroplumoidea** Gill, 1894
Family **Retroplumidae** Gill, 1894
Superfamily **Cancroidea** Latreille, 1802
Family **Atelecyclidae** Ortmann, 1893
**Cancridae** Latreille, 1802
**Cheiragonidae** Ortmann, 1893
**Corystidae** Samouelle, 1819
**Pirimelidae** Alcock, 1899
**Thiidae** Dana, 1852
Superfamily **Portunoidea** Rafinesque, 1815
Family **Geryonidae** Colosi, 1923
**Portunidae** Rafinesque, 1815

**Trichodactylidae** Milne Edwards, 1853
Superfamily **Bythograeoidea** Williams, 1980
Family **Bythograeidae** Williams, 1980
Superfamily **Xanthoidea** MacLeay, 1838
Family **Carpiliidae** Ortmann, 1893
**Eumedonidae** Dana, 1853
**Goneplacidae** MacLeay, 1838
**Hexapodidae** Miers, 1886
**Menippidae** Ortmann, 1893
**Panopeidae** Ortmann, 1893
**Pilumnidae** Samouelle, 1819
**Platyxanthidae** Guinot, 1977
**Pseudorhombilidae** Alcock, 1900
**Trapeziidae** Miers, 1886
**Xanthidae** MacLeay, 1838
Superfamily **Bellioidea** Dana, 1852
Family **Belliidae** Dana, 1852
Superfamily **Potamoidea** Ortmann, 1896
Family **Deckeniidae** Ortmann, 1897
**Platythelphusidae** Colosi, 1920
**Potamidae** Ortmann, 1896
**Potamonautidae** Bott, 1970
Superfamily **Pseudothelphusoidea** Ortmann, 1893
Family **Pseudothelphusidae** Ortmann, 1893
Superfamily **Gecarcinucoidea** Rathbun, 1904
Family **Gecarcinucidae** Rathbun, 1904
**Parathelphusidae** Alcock, 1910
Superfamily **Cryptochiroidea** Paulson, 1875
Family **Cryptochiridae** Paulson, 1875
Subsection **Thoracotremata** Guinot, 1977
Superfamily **Pinnotheroidea** de Haan, 1833
Family **Pinnotheridae** de Haan, 1833
Superfamily **Ocypodoidea** Rafinesque, 1815
Family **Camptandriidae** Stimpson, 1858
**Mictyridae** Dana, 1851
**Ocypodidae** Rafinesque, 1815
**Palicidae** Bouvier, 1898
Superfamily **Grapsoidea** MacLeay, 1838
Family **Gecarcinidae** MacLeay, 1838
**Glyptograpsidae** Schubart, Cuesta & Felder, 2001
**Grapsidae** MacLeay, 1838
**Plagusiidae** Dana, 1851
**Sesarmidae** Dana, 1851
**Varunidae** Milne Edwards, 1853

## LITERATURE CITED


Abele, L. G. 1987. Review of: Schram, F. R. 1986. *Crustacea.* Oxford University Press. *Journal of Crustacean Biology* 7:200–201.

———. 1991. Comparisons of morphological and molecular phylogeny of the Decapoda. In *Proceedings of the 1990 International Crustacean Conference,* ed. P. J. F. Davie and R. H. Quinn. *Memoirs of the Queensland Museum* 31:101–108.

Abele, L. G., and B. E. Felgenhauer. 1986. Phylogenetic and phenetic relationships among the lower Decapoda. *Journal of Crustacean Biology* 6:385–400.

Abele, L. G., W. Kim, and B. E. Felgenhauer. 1989. Molecular evidence for inclusion of the phylum Pentastomida in the Crustacea. *Molecular Biology and Evolution* 6:685–691.

Abele, L. G., and T. Spears. 1997. Issues and answers in the molecular phylogeny of the Crustacea. *Program and Abstracts, The Crustacean Society 1997 Summer Meeting, Mobile, Alabama*: 13.

Abele, L. G., T. Spears, and N. Cumberlidge. 1999. Biogeography and phylogeny of freshwater crabs based on molecular evidence. *Program and Abstracts, The Crustacean Society 1999 Summer Meeting, Lafayette, Louisiana*: 20.

Abele, L. G., T. Spears, W. Kim, and M. Applegate. 1992. Phylogeny of selected maxillopodan and other crustacean taxa based on 18S ribosomal nucleotide sequences: a preliminary analysis. *Acta Zoologica* 73: 271–392.

Aguinaldo, A. M., J. M. Turbeville, L. S. Linford, M. C. Rivera, J. R. Garey, R. A. Raffe, and J. A. Lake. 1997. Evidence for a clade of nematodes, arthropods, and other moulting animals. *Nature* 387:489–493.

Ahyong, S. T. 1997. Phylogenetic analysis of the Stomatopoda (Malacostraca). *Journal of Crustacean Biology* 17:695–715.

———. 2001. Revision of the Australian stomatopod Crustacea. *Records of the Australian Museum,* supplement 26:1–326.

Ahyong, S. T., and C. Harling. 2000. The phylogeny of the stomatopod Crustacea. *Australian Journal of Zoology* 48:607–642.

Akam, M. 1998. Hox genes in arthropod development and evolution. *Biological Bulletin* 195:373–374

Akam, M., M. Averof, J. Castelli-Gair, R. Dawes, F. Falciani, and D. Ferrier. 1994. The evolving role of Hox genes in arthropods. *Development* 1994(supplement):209–215.

Alcock, A. 1898. Materials for a carcinological fauna of India. No. 3. The Brachyura Cyclometopa. Part I. The Family Xanthidae. *Journal of the Asiatic Society of Bengal* 68(part 2, no. 1):67–233.

———. 1900. Materials for a carcinological fauna of India. No. 5. The Brachyura Primigenia or Dromiacea. *Journal of the Asiatic Society of Bengal* 68(part 2, no. 3):123–169.

Almeida, W. de O., and M. L. Christoffersen. 1999. A cladistic approach to relationships in Pentastomida. *Journal of Parasitology* 85:695–704.

Alonso, M. 1996. Crustacea Branchiopoda. *Fauna Iberica,* vol. 7. Madrid: Museo Nacional de Ciencias Naturales, 486 pp.

Amado, M. A. P., J.-S. Ho, and C. E. F. da Rocha. 1995. Phylogeny and biogeography of the Ergasilidae (Copepoda, Poecilostomatoidea), with reconsideration of the taxonomic status of the Vaigamidae. *Contributions to Zoology* 65:233–243.

Amoros, C. 1996. Branchiopodes II. Ordre des Cténopodes, Anomopodes, Onychopodes et Haplopodes (Ctenopoda Sars, 1865—Anomopoda Sars, 1865—Onychopoda Sars, 1865—Haplopoda Sars, 1865). In *Traité de Zoologie. Anatomie, Systématique, Biologie. Crustacés. Tome VII, Fascicule II. Généralités (suite) et Systématique,* ed. J. Forest, 353–383. Paris: Masson, 1002 pp.

Anderson, D. T. 1994. *Barnacles. Structure, function, development and evolution.* London: Chapman and Hall, 357 pp.

Andronov, V. N. 1974. Phylogenetic relations of large taxa within the suborder Calanoida (Crustacea, Copepoda). *Zoologischeskii Zhurnal* 53:1002–1012. [In Russian, with English summary]

Arbizu, P. M., and G. Moura. 1994. The phylogenetic position of the Cylindropsyllinae Sars (Copepoda, Harpacticoida) and the systematic status of the Leptopontiinae Lang. *Zoologisch Beiträge* 35:55–77.

Arhat, A., and T. C. Kaufman. 1999. Novel regulation of the homeotic gene *Scr* associated with a crustacean leg-to-maxilliped appendage transformation. *Development* 126:1121–1128.

Atkinson, R. J. A., and A. C. Taylor. 1988. Physiological ecology of burrowing decapods. In *Aspects of Decapod Crustacean Biology,* ed. A. A. Fincham and P. Rainbow. *Symposium of the Zoological Society of London* 59:201–226.

Avdeev, G. V., and B. I. Sirenko. 1991. Chitonophilidae fam. n., a new family of parasitic copepods from the chitons of the north-western Pacific. *Parazitologiya* 25:370–374.

Averof, M., and M. Akam. 1993. Hom/Hox genes of *Artemia*: implications for the origin of insect and crustacean body plans. *Current Biology* 3:73–78.

———. 1995a. Hox genes and the diversification of insect and crustacean body plans. *Nature* 376:420–423.

———. 1995b. Insect–crustacean relationships: insights from comparative developmental and molecular studies. *Philosophical Transactions of the Royal Society of London* 347B:293–303.

Averof, M., and N. H. Patel. 1997. Crustacean appendage evolution associated with changes in Hox gene expression. *Nature* 388:682–686.

Ax, P. 1999. *Das system der Metazoa II. Ein Lehrbuch der phylogenetischen Systematik.* Stuttgart: G. Fischer, 383 pp.

Baba, K. 1988. Chirostylid and galatheid crustaceans (Decapoda: Anomura) of the "Albatross" Philippine Expedition. *Researches on Crustacea,* special no. 2. Tokyo: The Carcinological Society of Japan, 203 pp.

Băcescu, M. 1972. *Archaeocuma* and *Schizocuma,* new genera of Cumacea from the American tropical waters. *Reue Roumaine de Biologie, serie Zoologie* 17: 241–245.

———. 1988. Cumacea I (Fam. Archaeocumatidae, Lampropidae, Bodotriidae, Leuconidae). In *Crustaceorum catalogus pars 7,* ed. H.-E. Gruner and L. B. Holthuis, 1–173. The Hague: SPB Academic Publishing.

———. 1992. Cumacea II (Fam. Nannastacidae, Diastylidae, Pseudocumatideae, Gynodiastylidae et Ceratocumatidae). In *Crustaceorum catalogus pars 8,* ed.

H.-E. Gruner and L. B. Holthuis, 1–468. The Hague: SPB Academic Publishing.

Băcescu, M., and I. Petrescu. 1999. Ordre des Cumacés (Cumacea Kroyer, 1846). In *Traité de Zoologie. Anatomie, Systématique, Biologie. Tome VII, Fascicule IIIA. Crustacés Péracarides,* ed. J. Forest. Memoires de l'Institut Oceanographique Fondation Albert Ier, Prince de Monaco, 19:391–428.

Baird, W. 1850. *The natural history of British Entomostraca.* London: The Ray Society, 364 pp.

Baker, A. de C., B. P. Boden, and E. Brinton. 1990. *A practical guide to the euphausiids of the world.* London: Natural History Museum Publications, The Natural History Museum, 96 pp.

Barber, P. H., and M. V. Erdmann. 2000. Molecular systematics of the Gonodactylidae (Stomatopoda) using mitochondrial cytochrome oxidase C (subunit 1) DNA sequence data. *Journal of Crustacean Biology,* special no. 2, 20:20–36.

Barnard, J. L., and J. Clark. 1984. Redescription of *Phoxocephalopsis zimmeri* with a new species, and establishment of the family Phoxocephalopsidae (Amphipoda) from Magellanic South America. *Journal of Crustacean Biology* 4:85–105.

Barnard, J. L., and M. M. Drummond. 1982. Gammaridean amphipoda of Australia, part V: superfamily Haustorioidea. *Smithsonian Contributions in Zoology* 360:1–148.

Barnard, J. L., and G. S. Karaman. 1982. Classificatory revisions in gammaridean Amphipoda (Crustacea), part 2. *Proceedings of the Biological Society of Washington* 95:167–187.

———. 1987. Revisions in classification of gammaridean Amphipoda (Crustacea), part 3. *Proceedings of the Biological Society of Washington* 100:856–875.

———. 1991. The families and genera of marine gammaridean Amphipoda (except marine gammaroids). *Records of the Australian Museum* Supplement 13(parts 1 and 2):1–866.

Barnard, J. L., and J. D. Thomas. 1988. Ipanemidae, new family, *Ipanema talpa,* new genus and species, from the surf zone of Brazil (Crustacea: Amphipoda: Haustorioidea). *Proceedings of the Biological Society of Washington* 101:614–621.

Barnes, R. D., and F. W. Harrison. 1992. Introduction. In *Microscopic anatomy of invertebrates,* vol. 9, *Crustacea,* ed. F. W. Harrison and A. G. Humes, 1–8. New York: Wiley-Liss, Inc.

Bate, C. S. 1862. *Catalogue of the specimens of amphipodous Crustacea in the collection of the British Museum.* London: British Museum of Natural History, iv + 399 pages, plates. 1–58.

Belk, D. 1996. Was sind "Urzeitkrebse"? *Stapfia 42, Zugleich Kataloge des O. Ö. Landesmuseums N. F.* 100:15–19.

Bellan-Santini, D. 1999. Ordre des Amphipodes (Amphipoda Latreille, 1816). In *Traité de Zoologie. Anatomie, Systématique, Biologie. Tome VII, Fascicule IIIA. Crustacés Péracarides,* ed. J. Forest. Memoires de l'Institut Oceanographique Fondation Albert Ier, Prince de Monaco, 19:93–176.

Bellan-Santini, D., and M. Ledoyer. 1986. Gammariens (Crustacea—Amphipoda) des Iles Marion et Prince Edward. *Boletino Museum Civico Storia Naturale Verona* 13:349–435.

Bellwood, O. 1996. A phylogenetic study of the Calappidae H. Milne Edwards 1837 (Crustacea: Brachyura) with a reappraisal of the status of the family. *Zoological Journal of the Linnean Society* 118:165–193.

———. 1998. Evolution and biogeography of the Calappidae (Decapoda: Brachyura). *Proceedings and Abstracts of the 4th International Crustacean Congress, Amsterdam*: 71 (abstract 101).

Benton, M. J. (editor). 1993. *The fossil record 2.* London: Chapman and Hall, 845 pp.

Berge, J., G. Boxshall, and W. Vader. 2000. Cladistic analysis of the Amphipoda. *Abstracts of the 10th Colloqium on Amphipoda* (http://www.odu.edu/%7Ejrh100f/amphome): 1.

Berge, J., W. Vader, and C. O. Coleman. 1998. Cladistic analysis of the amphipod families Odiidae and Ochlesidae. *Proceedings and Abstracts of the 4th International Crustacean Congress, Amsterdam*: 53 (abstract 40).

———. 1999. A cladistic analysis of the amphipod families Ochlesidae and Odiidae, with description of a new species and genus. In *Crustaceans and the biodiversity crisis. Proceedings of the 4th International Crustacean Congress, Amsterdam,* vol. 1, ed. F. R. Schram and J. C. von Vaupel Klein, 239–265. Leiden: Brill.

Bergström, J. 1992. The oldest arthropods and the origin of the Crustacea. *Acta Zoologica* 73:287–292.

Boore, J. L., T. M. Collins, D. Stanton, L. L. Daehler, and W. M. Brown. 1995. Deducing arthropod phylogeny from mitochondrial DNA rearrangements. *Nature* 376:163–165.

Boore, J. L., D. Lavrov, and W. M. Brown. 1998. Gene translocation links insects and crustaceans. *Nature* 392:667–668.

Bott, R. 1970a. Die Süsswasserkrabben von Europa, Asien, Australien und ihre Stammesgeschichte. Eine Revision der Potamoidea und der Parathelphusoidea (Crustacea, Decapoda). *Abhandlungen der Senckenbergischen Naturforschenden Gesellschaft* 526:1–338.

———. 1970b. Betrachtungen uber die Entwicklungsgeschichte und Verbreitung der Süsswasser-krabben nach der Sammlung des Naturhistorischen Museums in Genf/Schweitz. *Revue Suisse de Zoologie* 77:327–344.

Bourne, G. C. 1922. The Raninidae: a study in carcinology. *Journal of the Linnean Society of London (Zoology)* 35:25–79.

Bousfield, E. L. 1982a. Amphipoda. Gammaridea and Ingolfiellidea. In *Synopsis and classification of living organisms,* vol. 2, ed. S. B. Parker, 254–284, 293–294. New York: McGraw-Hill.

———. 1982b. Amphipoda (palaeohistory). In *McGraw-Hill Yearbook of Science & Technology 1982–1983,* 96–100. New York: McGraw-Hill.

———. 1983. An updated phyletic classification and palaeohistory of the Amphipoda. In *Crustacean Issues 1. Crustacean Phylogeny,* ed. F. R. Schram, 257–277. Rotterdam: A. A. Balkema Press, 372 pp.

———. 2000a. An updated commentary on the phyletic classification of amphipod crustaceans. *Abstracts of the 10th Colloqium on Amphipoda* (http://www.odu.edu/%7Ejrh100f/amphome): 1.

———. 2000b. The value of phyletic classification in biogeographical analyses. *Abstracts of the 10th Colloqium on Amphipoda* (http://www.odu.edu/%7Ejrh100f/amphome): 1–2.

———. 2001. An updated commentary on phyletic classification of the amphipod Crustacea and its applicability to the North American fauna. *Amphipacifica* 3:49–120.

Bousfield, E. L., and K. E. Conlan. 1990. Crustaceans. In *The new Encyclopaedia Britannica,* 15th edition, vol. 16, 840–859. Chicago: Encyclopaedia Brittanica, Inc.

Bousfield, E. L., and E. A. Hendrycks. 1994. A revision of the family Pleustidae (Crustacea: Amphipoda: Leucothoidea). Systematics and biogeography of component subfamilies. Part I. *Amphipacifica* 1:17–58.

———. 1997. The amphipod superfamily Eusiroidea in the North American Pacific region. II. Family Calliopidae. Systematics and distributional ecology. *Amphipacifica* 2:3–66.

Bousfield, E. L., and P. M. Hoover. 1995. The amphipod superfamily Pontoporeioidea on the Pacific coast of North America. II. Family Haustoriidae. Genus *Eohaustorius* J. L. Barnard: systematics and distributional ecology. *Amphipacifica* 2:35–64.

———. 1997. The amphipod superfamily Corophioidea on the Pacific coast of North America. Part V. Family Corophiidae. Corophinae, new subfamily. Systematics and distributional ecology. *Amphipacifica* 2: 67–139.

Bousfield, E. L., and J. A. Kendall. 1994. The amphipod superfamily Dexaminoidea on the North American Pacific coast; families Atylidae and Dexaminidae: systematics and distributional ecology. *Amphipacifica* 1:3–66.

Bousfield, E. L., and C.-T. Shih. 1994. The phyletic classification of amphipod crustaceans: problems in resolution. *Amphipacifica* 1:76–134.

Boutin, C., and M. Missouli. 1988. *Metacrangonyx gineti,* n. sp., d'une source du haut-Atlas Marocain det la famille des Metacrangonyctidae, n. fam. *Vie et Milieu* 38:67–84.

Bower, B. 1999. DNA's evolutionary dilemma: genetic studies collide with the mystery of human evolution. *Science News* 155:88–90.

Bowman, T. E., and L. G. Abele. 1982. Classification of the Recent Crustacea. In *Systematics, the fossil record, and biogeography,* ed. L. G. Abele, 1–27, vol. I of *The biology of Crustacea,* ed. D. E. Bliss. New York: Academic Press.

Bowman, T. E., S. P. Garner, R. R. Hessler, T. M. Iliffe, and H. L. Sanders. 1985. Mictacea, a new order of Crustacea Peracarida. *Journal of Crustacean Biology* 5:74–78.

Bowman, T. E., and H. E. Gruner. 1973. The families and genera of Hyperiidea. *Smithsonian Contributions to Zoology* 146:1–64.

Bowman, T. E., and T. Iliffe. 1985. *Mictocaris halope,* a new unusual peracaridan crustacean from marine caves on Bermuda. *Journal of Crustacean Biology* 5: 58–73.

Boxshall, G. A. 1983. A comparative functional analysis of the major maxillopodan groups. In *Crustacean Issues 1. Crustacean Phylogeny,* ed. F. R. Schram, 121–143. Rotterdam: A. A. Balkema Press, 372 pp.

———. 1986. Panel discussion: copepod phylogeny. *Syllogeus* 58:197–208.

———. 1988. A review of the copepod endoparasites of brittle stars (Ophiuroida). *Bulletin of the British Museum of Natural History (Zoology)* 54:261–270.

———. 1991. A review of the biology and phylogenetic relationships of the Tantulocarida, a subclass of Crustacea recognized in 1983. *Verhandelungen der Deutschen Zoologischen Gesellschaft* 84:271–279.

———. 1992. Synopsis of group discussion on the Maxillopoda. *Acta Zoologica* 73:335–337.

———. 1996. Classe de Tantulocarides (Tantulocarida Boxshall et Lincoln, 1983). In *Traité de Zoologie. Anatomie, Systématique, Biologie. Crustacés. Tome VII, Fascicule II. Généralités (suite) et Systématique,* ed. J. Forest, 399–408. Paris: Masson, 1002 pp.

———. 1999. Ordre des Spélaeogriphacés (Spelaeogriphacea Gordon, 1957). In *Traité de Zoologie. Anatomie, Systématique, Biologie. Tome VII, Fascicule IIIA. Crustacés Péracarides,* ed. J. Forest. Mémoires de l'Institut Océanographique Fondation Albert, Ier, Prince de Monaco, 19:35–38.

Boxshall, G. A., and D. Defaye. 1996. Classe des Mystacocarides (Mystacocarida Pennak et Zinn, 1943). In *Traité de Zoologie. Anatomie, Systématique, Biologie. Crustacés. Tome VII, Fascicule II. Généralités (suite) et Systématique,* ed. J. Forest, 409–424. Paris: Masson, 1002 pp.

Boxshall, G. A., F. D. Ferrari, and H. Tiemann. 1984. The ancestral copepod: towards a consensus of opinion at the First International Conference on Copepoda. *Crustaceana* 7(supplement):68–84.

Boxshall, G. A., and R. Huys. 1989a. New tantulocarid, *Stygotantulus stocki,* parasitic on harpacticoid copepods, with an analysis of the phylogenetic relationships within the Maxillopoda. *Journal of Crustacean Biology* 9:126–140.

———. 1989b. New family of deep-sea plankonic copepods, the Paralubbockiidae (Copepoda: Poecilostomatoida). *Biological Oceanography* 6:163–173.

Boxshall, G. A., and D. Jaume. 1999. On the origin of misophrioid copepods from anchialine caves. *Crustaceana* 72:957–963.

———. 2000. Discoveries of cave misophrioids (Crustacea: Copepoda) shed new light on the origin of anchialine faunas. *Zoologische Anzeiger* 239:1–19.

Boxshall, G. A., and R. J. Lincoln. 1983. Tantulocarida, a new class of Crustacea ectoparasitic on other crustaceans. *Journal of Crustacean Biology* 3:1–16.

———. 1987. The life cycle of the Tantulocarida (Crustacea). *Philosophical Transactions of the Royal Society of London* 315B:267–303.

Boxshall, G. A., and S. Ohtsuka. 2001. Two new families of copepods (Copepoda: Siphonostomatoida) parasitic on echinoderms. *Journal of Crustacean Biology* 21:96–105.

Boxshall, G. A., J.-O. Strömberg, and E. Dahl (editors). 1992. The Crustacea: origin and evolution. *Acta Zoologica* 73:271–392.

Brady, G. S., and A. M. Norman, 1889. A monograph on the marine and freshwater Ostracoda of the North Atlantic and of northwestern Europe. Section I: Podocopa. *Scientific Transactions of the Royal Dublin Society,* series 2, 4:63–270.

Braga, E., R. Zardoya, A. Meyer, and J. Yen. 1999. Mitochondrial and nuclear rRNA based copepod phylogeny with emphasis on the Euchaetidae (Calanoida). *Marine Biology* 133:79–90.

Brandt, A., J. A. Crame, H. Polz, and M. R. A. Thomson. 1999. Late Jurassic Tethyan ancestry of Recent southern high-latitude marine isopods (Crustacea, Malacostraca). *Palaeontology* 42:663–675.

Brasil-Lima, I. M. 1998. Malacostraca—Peracarida. Isopoda. Epicaridea. In *Catalogue of Crustacea of Bra-*

*zil,* ed. P. S. Young, 635–644. Rio de Janeiro: Museu Nacional.

Briggs, D. E. G. 1983. Affinities and early evolution of the Crustacea: the evidence of the Cambrian fossils. In *Crustacean issues 1. Crustacean phylogeny,* ed. F. R. Schram, 1–22. Rotterdam: A. A. Balkema Press, 372 pp.

Briggs, D. E. G., D. H. Erwin, and F. J. Collier. 1994. *The fossils of the Burgess Shale.* Washington, D.C.: Smithsonian Institution Press, 238 pp.

Briggs, D. E. G., and R. A. Fortey. 1989. The early radiation and relationships of the major arthropod groups. *Science* 246:241–243.

———. 1992. Chapter 10. The Early Cambrian radiation of arthropods. In *Origin and early evolution of the Metazoa,* ed. J. H. Lipps and P. W. Signor, 335–373. New York: Plenum Press.

Briggs, D. E. G., R. A. Fortey, and M. A. Wills. 1992. Morphological disparity in the Cambrian. *Science* 256:1670–1673.

———. 1993a. How big was the Cambrian explosion? A taxonomic and morphologic comparison of Cambrian and Recent arthropods. In *Evolutionary patterns and processes, Linnean Society Symposium,* ed. D. R. Lees and D. Edwards, 33–44. London: Linnean Society.

Briggs, D. E. G., M. J. Weedon, and M. A. Whyte. 1993b. Arthropoda (Crustacea excluding Ostracoda). In *The Fossil Record 2,* ed. M. J. Benton, 321–342. London: Chapman and Hall.

Briggs, D. E. G., and H. B. Whittington. 1981. Relationships of arthropods from the Burgess Shale and other Cambrian sequences. In *Short Papers for the Second International Symposium on the Cambrian System,* ed. M. E. Taylor, 38–41. U.S. Geological Survey Open-File Report 81-743:1–254.

Brower, A. V. Z., R. DeSalle, and A. Vogler. 1996. Gene trees, species trees, and systematics: a cladistic perspective. *Annual Review of Ecology and Systematics* 27:423–450.

Brtek, J. 1964. Eine neue gattung und familie der ordnung Anostraca. *Annotationes Zoologicae et Botanicae (Slovenské Národné Múzeum)* 7:1–7.

———. 1995. Some notes on the taxonomy of the family Chirocephalidae (Crustacea, Branchiopoda, Anostraca). *Zborník Slovenského Národného Múzea (Prirodné Vedy)* 41:3–15.

———. 1997. Checklist of the valid and invalid names of the “large branchiopods” (Anostraca, Notostraca, Spinicaudata, and Laevicaudata), with a survey of the taxonomy of all Branchiopoda. *Zborník Slovenského Národného Múzea (Prírondné Vedy)* 43:1–66.

Brtek, J., and G. Mura. 2000. Revised key to families and genera of the Anostraca with notes on their geographical distribution. *Crustaceana* 73:1037–1088.

Brtek, J., and A. Thiéry. 1995. The geographic distribution of the European branchiopods (Anostraca, Notostraca, Spinicaudata, Laevicaudata). *Hydrobiologia* 298: 263–280.

Bruce, A. J. 1993. *Kakadukaris glabra* gen. nov., spec. nov., a new freshwater shrimp from the Kakadu National Park, Northern Territory, Australia (Crustacea: Decapoda: Palaemonidae) with designation of a new subfamily Kakaducaridinae. *Hydrobiologia* 268:27–44.

Bruce, N. L. 1984. A new family for the isopod crustacean genus *Tridentella* Richardson, 1905, with description of a new species from Fiji. *Zoological Journal of the Linnean Society* 80:447–455.

———. 1988. *Hadromastax merga,* a new genus and species of marine isopod crustacean (Limnoriidae) from southeastern Australia, with discussion on the status of the families Keuphyliidae and Lynseiidae. *Proceedings of the Biological Society of Washington* 101:346–353.

———. 1993. Two new genera of marine isopod crustaceans (Flabellifera: Sphaeromatidae) from southern Australia, with a reappraisal of the Sphaeromatidae. *Invertebrate Taxonomy* 7:151–171.

Bruce, N. L., and H.-G. Müller. 1991. A new family for the isopod genus *Hadromastax* Bruce, 1988, with a description of a new species from the Society Islands. *Zoological Journal of the Linnean Society* 101:51–58.

Brünnich, M. Th. 1772. *Zoologiae fundamenta praelectionibus academicis accomodata. Grunde i Dyrelaeren.* Hafniae et Lipsiae [= Copenhagen and Leipzig]: Apud Frider. Christ. Pelt., 254 pp. [not seen; as cited in Holthuis, 1991].

Brusca, R. C. 1990. Ferrara, F. (editor). *Proceedings of the second symposium on the biology of terrrestrial isopods* (Review). *Journal of Crustacean Biology* 10: 568–570.

———. 2000. Unraveling the history of arthropod biodiversification. In *Our unknown planet: recent discoveries and the future. Proceedings of the 45th Annual Systematics Symposium of the Missouri Botanical Garden. Annals of the Missouri Botanical Garden* 87:13–25.

Brusca, R. C., and G. J. Brusca. 1990. *Invertebrates.* Sunderland, Massachusetts: Sinauer Associates, Inc., 922 pp.

Brusca, R. C., and G. D. F. Wilson. 1991. A phylogenetic analysis of the Isopoda with some classificatory recommendations. In *Proceedings of the 1990 International Crustacean Conference,* ed. P. J. F. Davie and R. H. Quinn. *Memoirs of the Queensland Museum* 31:143–204.

Buckeridge, J. S. 1983. Fossil barnacles (Cirripedia: Thoracica) of New Zealand and Australia. *New Zealand Geological Survey Paleontological Bulletin* 50:1–151.

———. 1995. Phylogeny and biogeography of the primitive Sessilia and a consideration of a Tethyan origin for the group. In *Crustacean Issues 10. New Frontiers in Barnacle Evolution,* ed. F. R. Schram and J. T. Høeg, 255–267. Rotterdam: A. A. Balkema Press, 318 pp.

Burkenroad, M. D. 1963. The evolution of the Eucarida (Crustacea, Eumalacostraca) in relation to the fossil record. *Tulane Studies in Geology* 2:3–16.

———. 1981. The higher taxonomy and evolution of Decapoda (Crustacea). *Transactions of the San Diego Society of Natural History* 19:251–268.

Butterfield, N. J. 1994. Burgess Shale-type fossils from a Lower Cambrian shallow-shelf sequence in northwestern Canada. *Nature* 369:477–479.

Cals, Ph. 1996. Classe des Rémipèdes (Remipedia Yager, 1981). In *Traité de Zoologie. Anatomie, Systématique, Biologie. Crustacés. Tome VII, Fascicule II. Généralités (suite) et Systématique,* ed. J. Forest, 385–397. Paris: Masson, 1002 pp.

Cals, Ph., and Th. Monod. 1988. Évolution et biogéographie des Crustacés Thermosbénacés. *Comptes Ren-*

*dus Hebdomadaires des Seances de l'Academie des Sciences,* series 3, 307:341–348.

Camp, D. K., W. G. Lyons, and T. H. Perkins. 1998. Checklists of selected shallow-water marine invertebrates of Florida. *Florida Marine Research Institute Technical Report TR-3, Florida Department of Environmental Protection,* 1998:i–xv, 1–238.

Cantino, P. D. 2000. Phylogenetic nomenclature: addressing some concerns. *Taxon* 49:85–93.

Cantino, P. D., H. N. Bryant, K. de Queiroz, M. J. Donoghue, T. Eriksson, D. M. Hillis, and M. S. Y. Lee. 1999. Species names in phylogenetic nomenclature. *Systematic Biology* 48:790–807.

Cappola, V. A. 1999. Phylogeny of stomatopod crustaceans. *Program and Abstracts, The Crustacean Society 1999 Summer Meeting, Lafayette, Louisiana:* 25.

Cappola, V. A., and R. B. Manning. 1998. Stomatopod superfamily Gonodactyloidea is not a natural group. *Proceedings and Abstracts of the 4th International Crustacean Congress, Amsterdam:* 77 (abstract 121).

Carpenter, J. H. 1999. Behavior and ecology of *Speleonectes epilimnius* (Remipedia, Speleonectidae) from surface water of an anchialine cave on San Salvador Island, Bahamas. *Crustaceana* 72:979–991.

Carroll, S. B. 1995. Homeotic genes and the evolution of arthropods and chordates. *Nature* 376:479–485.

Casanova, J.-P., L. De Jong, and E. Faure. 1998. Interrelationships of the two families constituting the Lophogastrida (Crustacea: Mysidacea) inferred from morphological and molecular data. *Marine Biology* 132:59–65.

Castro, P. 2000. Crustacea Decapoda: A revision of the Indo-West Pacific species of palicid crabs (Brachyura Palicidae). In *Résultats des Campagnes Musorstom,* vol. 21, ed. A. Crosnier. *Mémoires du Museum National d'Histoire Naturelle,* vol. 184, 437–610.

Chace, F. A. 1992. On the classification of the Caridea (Decapoda). *Crustaceana* 63:70–80.

———. 1997. The caridean shrimps (Crustacea: Decapoda) of the *Albatross* Philippine Expedition, 1907–1910, Part 7: families Atyidae, Eugonatonotidae, Rhynchocinetidae, Bathypalaemonellidae, Processidae, and Hipploytidae. *Smithsonian Contributions to Zoology* 587:1–106.

Chappuis, P. A. 1915. *Bathynella natans* und ihre Stellung im System. *Zoologisches Jahrbüch der Systematik* 40:147–176.

Chia, O. K. S., and P. K. L. Ng. 1998. Is the Sundathelphusidae Bott, 1969 a valid taxon? A cladistic appraisal. *Proceedings and Abstracts of the 4th International Crustacean Congress, Amsterdam*: 72 (abstract 103).

Chia, D. G. B., and P. K. L. Ng. 2000. A revision of *Eumedonus* H. Milne Edwards, 1834 and *Gonatonotus* White, 1847 (Crustacea: Decapoda: Brachyura: Eumedonidae), two genera of crabs symbiotic with sea urchins. *Journal of Natural History* 34:15–56.

Chow, S., M. Okazaki, M. Takeda, and T. Kubota. 2000. A rare abyssal shrimp, *Galatheocaris abyssalis,* found in the stomach of a lancetfish. *Crustaceana* 73:243–246.

Christoffersen, M. L. 1986. Phylogenetic relationships between Oplophoridae, Atyidae, Pasiphaeidae, Alvinocarididae fam. n., Bresiliidae, Psalidopodidae and Disciadidae (Crustacea Caridea Atyoidea). *Bolletim de Zoologie* (University Sao Paulo, Brazil) 10:273–281.

———. 1987. Phylogenetic relationships of hippolytid genera, with an assignment of new families for the Crangonoidea and Alpheoidea (Crustacea, Decapoda, Caridea). *Cladistics* 3:348–362.

———. 1988a. Phylogenetic systematics of the Eucarida (Crustacea Malacostraca). *Revista Brasiliera de Zoologia* 5:325–351.

———. 1988b. Genealogy and phylogenetic classification of the world Crangonidae (Crustacea, Caridea), with a new species and new records for the southwestern Atlantic. *Revista Nordestina de Biologia* 6:43–59.

———. 1989. Phylogeny and classification of the Pandaloidea (Crustacea, Caridea). *Cladistics* 5:259–274.

———. 1990. A new superfamily classification of the Caridea (Crustacea: Pleocyemata) based on phylogenetic pattern. *Zeitschrift für Zoologische Systematik und Evolutionsforschung* 28:94–106.

———. 1994. An overview of cladistic applications. *Revista Nordestina de Biologia* 9:133–141.

———. 1998. Malacostraca. Eucarida. Caridea. Crangonidea and Alpheoidea (except Glyphocrangoidae and Crangonidae). In *Catalogue of Crustacea of Brazil,* ed. P. S. Young, 351–372. Rio de Janeiro: Museu Nacional.

Clark, P. F., and W. R. Webber. 1991. A redescription of *Macrocheira kaempferi* (Temminck, 1836) zoeas with a discussion of the classification of the Majoidea Samouelle, 1819 (Crustacea: Brachyura). *Journal of Natural History* 25:1259–1279.

Coelho, P. A., and P. A. Coelho Filhol. 1993. Proposta de classificação da família Xanthidae (Crustacea, Decapoda, Brachyura) através da taxonomia numérica. *Revista Brasileira de Zoologia* 10:559–580.

Cohen, A. C. 1982. Ostracoda. In *Synopsis and classification of living organisms,* ed. S. Parker, 181–202. New York: McGraw-Hill.

Cohen, A. C., J. W. Martin, and L. S. Kornicker. 1998. Homology of Holocene ostracode biramous appendages with those of other crustaceans: the protopod, epipod, exopod, and endopod. *Lethaia* 31:251–265.

Cohen, B. F. 1998. Dendrotiidae (Crustacea: Isopoda) of the southeastern Australian continental slope. *Memoirs of the Museum of Victoria* 57:1–38.

Coineau, N. 1996. Sous-classe des Eumalacostracés (Eumalacostraca Grobben, 1892) super-ordre des Syncarides (Syncarida Packard, 1885). In *Traité de Zoologie. Anatomie, Systématique, Biologie. Crustacés. Tome VII, Fascicule II. Généralités (suite) et Systématique,* ed. J. Forest, 897–954. Paris: Masson, 1002 pp.

Coleman, C. O., and J. L. Barnard. 1991. Revision of Iphimediidae and similar families (Amphipoda: Gammaridea). *Proceedings of the Biological Society of Washington* 104:253–268.

Colgan, D. J., A. McLauchlan, G. D. F. Wilson, S. P. Livingston, G. D. Edgecomb, J. Macaranas, G. Cassis, and M. R. Gray. 1998. Histone H3 and U2 snRNA DNA sequences and arthropod molecular evolution. *Australian Journal of Zoology* 46:419–437.

Cookson, L. J., and G. C. B. Poore. 1994. New species of *Lynseia* and transfer of the genus to Limnoriidae. *Memoirs of the Museum of Victoria* 54:179–189.

Crandall, K. A. 1999. On the origin(s) of freshwater crayfishes. *Program and Abstracts, The Crustacean Society 1999 Summer Meeting, Lafayette, Louisiana*: 27.

Crandall, K. A., D. J. Harris, and J. W. Fetzner, Jr. 2000. The monophyletic origin of freshwater crayfish estimated from nuclear and mitochondrial DNA sequences. *Proceedings of the Royal Society of London* B 267:1679–1686.

Crease, T. J., and D. J. Taylor. 1998. The origin and evolution of variable-region helices in V4 and V7 of the small-subunit ribosomal RNA of branchiopod crustaceans. *Molecular Biology and Evolution* 15:1430–1446.

Cuesta, J. A., and C. D. Schubart. 1999. Proposed classification of the Grapsidae and Gecarcinidae (Decapoda, Brachyura) on the basis of larval morphology. *Program and Abstracts, The Crustacean Society 1999 Summer Meeting, Lafayette, Louisiana*: 52.

Cuesta, J. A., C. D. Schubart, and A. Rodríguez. 2000. Larval development of *Brachynotus sexdentatus* (Risso, 1827) (Decapoda, Brachyura) reared under laboratory conditions, with notes on larval characters of the Varunidae. *Invertebrate Reproduction and Development* 38:207–223.

Cumberlidge, N. 1987. Notes on the taxonomy of West African gecarcinucids of the genus *Globonautes* Bott, 1959 (Decapoda, Brachyura). *Canadian Journal of Zoology* 65:2210–2215.

———. 1991. The respiratory system of *Globonautes macropus* (Rathbun, 1898), a terrestrial freshwater crab from Liberia (Gecarcinucoidea, Gecarcinucidae). *Crustaceana* 61:69–80.

———. 1996a. A taxonomic revision of freshwater crabs (Brachyura, Potamoidea, Gecarcinucidae) from the Upper Guinea forest of West Africa. *Crustaceana* 69: 681–695.

———. 1996b. On the Globonautinae Bott, 1969, freshwater crabs from West Africa (Brachyura: Potamoidea: Gecarcinucidae). *Crustaceana* 69:809–820.

———. 1999. *The freshwater crabs of West Africa. Family Potamonautidae.* Paris: IRD Éditions, Faune et Flore Tropicales 35, 1–382.

Cumberlidge, N., and R. Sachs. 1991. Ecology, distribution, and growth in *Globonautes macropus* (Rathbun, 1898), a tree-living freshwater crab from the rain forests of Liberia (Parathelphusoidea, Gecarcinucidae). *Crustaceana* 61:55–68.

Cumberlidge, N., and R. von Sternberg. 1999. Phylogenetic relationships of the freshwater crabs of Lake Tanganyika (Decapoda, Brachyura). In *Crustaceans and the biodiversity crisis. Proceedings of the 4th International Crustacean Congress, Amsterdam,* vol. 1, ed. F. R. Schram and J. C. von Vaupel Klein, 405–422. Leiden: Brill.

Cumberlidge, N., R. von Sternberg, I. R. Bills, and H. Martin. 1999. A revision of the genus *Platythelphusa* A. Milne-Edwards, 1887 from Lake Tanganyika, East Africa (Decapoda: Potamoidea: Platythelphusidae). *Journal of Natural History* 33:1487–1512.

Cunningham, C. W., N. W. Blackstone, and L. W. Buss. 1992. Evolution of king crabs from hermit crab ancestors. *Nature* 355:539–542.

Dahl, E. 1983a. Alternatives in malacostracan evolution. *Australian Museum Memoir* 18:1–5.

———. 1983b. Malacostracan phylogeny and evolution. In *Crustacean issues 1. Crustacean phylogeny,* ed. F. R. Schram, 189–212. Rotterdam: A. A. Balkema Press, 372 pp.

———. 1987. Malacostraca maltreated—the case of the Phyllocarida. *Journal of Crustacean Biology* 7:721–726.

———. 1991. Crustacea Phyllopoda and Malacostraca: a reappraisal of cephalic dorsal shield and fold systems. *Philosophical Transactions of the Royal Society of London* 334B:1–26.

———. 1992. Aspects of malacostracan evolution. *Acta Zoologica* 73:339–346.

Dahl, E., and J.-O. Strömberg. 1992. Introduction: discovering crustacean diversity. *Acta Zoologica* 73: 271–272.

Dahl, E., and J.-W. Wägele. 1996. Sous-classe des Phyllocarides (Phyllocarida Packard, 1879). In *Traité de Zoologie. Anatomie, Systématique, Biologie. Crustacés. Tome VII, Fascicule II. Généralités (suite) et Systématique,* ed. J. Forest, 865–896. Paris: Masson, 1002 pp.

Dahms, H.-U. 1990. Naupliar development of Harpacticoida (Crustacea, Copepoda) and its significance for phylogenetic systematics. *Microfauna Marina* 6: 169–272.

———. 1993. Copepodid development in Harpacticoida (Crustacea, Copepoda). *Microfauna Marina* 8:195–245.

Dai, A.-Y. 1997. A revision of freshwater crabs of the genus *Nanhaipotamon* Bott, 1968 from China (Crustacea: Decapoda: Brachyura: Potamidae). *Raffles Bulletin of Zoology* 45:209–235.

Dai, A.-Y., and M. Türkay. 1997. Revision of the Chinese freshwater crabs previously placed in the genus *Isolapotamon* Bott, 1968 (Crustacea: Decapoda: Brachyura: Potamidae). *Raffles Bulletin of Zoology* 45: 237–264.

Dai, A.-Y., and S.-I. Yang. 1991. *Crabs of the China Seas.* Beijing: China Ocean Press, and Berlin: Springer Verlag, 682 pp. [1991 English reprint of a 1984 Chinese volume]

Dai, A.-Y., X. M. Zhou, and W. D. Peng. 1995. Eight new species of the genus *Sinopotamon* from Jiangxi Province, China (Crustacea, Decapoda, Brachyura, Potamidae). *Beaufortia* 45:61–76.

Damkaer, D. M. 1996. Copepod taxonomy: discovery vs. recognition. *Proceedings of the Biological Society of Washington* 109:687–694.

Dana, J. D. 1849. Conpsectus crustaceorum quae in orbis terrarum circumnavigatione, Carolo Wilkes e classe Reipublicae, Foederate Duce, lexit et descripsit (continued). *American Journal of Sciences and Arts* 8(2): 424–428.

———. 1852. On the classification of the Crustacea Choristopoda or Tetradecapoda. *American Journal of Sciences and Arts* 14(2):297–316.

Danielopol, D. L., and J. M. Betsch. 1980. Ostracodes terrestres de Madagascar: systematique, origine, adaptations. *Revue d'Ecologie et de Biologie du Sol* 17: 87–123.

da Rocha, C. E. F., and T. M. Iliffe. 1991. Speleoithonidae, a new family of Copepoda (Cyclopoida) from anchialine caves in the Bahama Islands. *Sarsia* 76:167–175.

Davidson, E. H., K. J. Peterson, and R. A. Cameron. 1995. Origin of bilaterian body plans: evolution of developmental regulatory mechanisms. *Science* 270: 1319–1325.

Davie, P. J. F. 1990. A new genus and species of marine crayfish, *Palibythus magnificus,* and new records of *Palinurellus* (Decapoda: Palinuridae) from the Pacific Ocean. *Invertebrate Taxonomy* 4:685–695.

Day, J. 1980. Southern African Cumacea, part 4. Families Gynodiastylidae and Diastylidae. *Annals of the South African Museum* 82:187–292.

De Broyer, C., and K. Jazdzewski. 1993. Contribution to

the marine biodiversity inventory. A checklist of the Amphipoda (Crustacea) of the Southern Ocean. *Documents de Travail de l'Institut Royal des Sciences Naturelles des Belgique* 73:1–154.

De Jong-Moreau, L., and J.-P. Casanova. 2001. The foreguts of the primitive families of the Mysida (Crustacea, Peracarida): a transitional link between those of the Lophogastrida (Crustacea, Mysidacea) and the most evolved Mysida. *Acta Zoologica* 82:137–147.

Delle Cave, L., and A. M. Simonetta. 1991. Early Paleozoic arthropods and problems of arthropod phylogeny, with some notes on taxa of doubtful affinities. In *The early evolution of Metazoa and the significance of problematic taxa,* ed. A. M. Simonetta and S. Conway Morris, 189–244. London: Cambridge University Press.

de Queiroz, K., and J. Gauthier. 1994. Toward a phylogenetic system of biological nomenclature. *Trends in Ecology and Evolution* 9:27–31.

Diesing, K. M. 1835 (1836). Versuch einter Monographie der Gattung Pentastoma. *Annalen des Wiener Museums der Naturgeschichte* 1:1–32.

Doyle, J. J. 1997. Trees within trees: genes and species, molecules and morphology. *Systematic Biology* 46: 537–553.

Drach, P., and D. Guinot. 1983. Les Inachoididae Dana, famille de Majoidea caractérisée par des connexions morphologiques d'un type nouveau entre carapace, pleurites, sternites, et pléon. *Comptes Rendus Hebdomadaires des Seances de l'Academie des Sciences,* series 3, 297:37–42.

Dupuis, C. 1975. Objections aux propositions de Bousfield & Holthuis (1969) concernant une douzaine de genres d'Amphipodes. Z.N.(S.) 1879. *Bulletin of Zoological Nomenclature* 32:3–5.

Edgecombe, G. D. (editor). 1998. *Arthropod fossils and phylogeny.* New York: Columbia University Press, 347 pp.

Edgecomb, G. D., G. D. F. Wilson, D. J. Colgan, M. R. Gray, and G. Cassis. 2000. Arthropod cladistics: combined analysis of histone H3 and U2 snRNA sequences and morphology. *Cladistics* 16:155–203.

Eernisse, D. J. 1997. Arthropod and annelid relationships re-examined. In *Arthropod relationships,* ed. R. A. Fortey and R. H. Thomas, 43–56. Systematics Association Special Volume Series 55. London: Chapman and Hall.

———. 1998. A brief guide to phylogenetic software. *Trends in Genetics* 14:473–475.

Elofsson, R. 1992. To the question of eyes in primitive crustaceans. *Acta Zoologica* 73:369–372.

Emerson, M. J., and F. R. Schram. 1990. The origin of crustacean biramous appendages and the evolution of Arthropoda. *Science* 250:667–669.

———. 1991. Remipedia, part 2, paleontology. *Proceedings of the San Diego Society of Natural History* 7: 1–52.

———. 1997. Theories, patterns, and reality: game plan for arthropod phylogeny. In *Arthropod relationships,* ed. R. A. Fortey and R. H. Thomas, 67–86. Systematics Association Special Volume series 55. London: Chapman and Hall.

Erhard, F. 1995. Vergleichend- und funktionell-anatomische Untersuchungen am Pleon der Oniscidea (Crustacea, Isopoda) Zugleich ein Beitrag zu phylogenetischen Systematik der Landasseln. *Zoologia* 48:114 pp.

Fain, A. 1961. Les pentastomides de l'Afrique centrale. *Annales de Musee Royale de l'Afrique Centrale,* serie 8, 92:1–115.

Felder, D. L., J. W. Martin, and J. W. Goy. 1985. Patterns in early postlarval development of decapods. In *Crustacean issues 2. Larval growth,* ed. A. M. Wenner, 163–225. Rotterdam: A. A. Balkema Press, 236 pp.

Feldmann, R. M., and C. S. Hopkins. 1999. Reconsideration of the family status of fossil and extant calappid crabs. *Program and Abstracts, The Crustacean Society 1999 Summer Meeting, Lafayette, Louisiana*: 29.

Felgenhauer, B. E., and L. G. Abele. 1983. Phylogenetic relationships among the shrimp-like decapods (Penaeoidea, Caridea, Stenopodidea). In *Crustacean issues 1. Crustacean phylogeny,* ed. F. R. Schram, 291–311. Rotterdam: A. A. Balkema Press, 372 pp.

Felgenhauer, B. E., L. G. Abele, and D. L. Felder. 1992. Remipedia. In *Microscopic anatomy of invertebrates,* vol. 9, *Crustacea,* ed. F. W. Harrison and A. G. Humes, 225–247. New York: Wiley-Liss.

Ferrara, F. (editor). 1989. Proceedings of the second symposium on the biology of terrestrial isopods. *Monitore Zoologico Italiano, n.s., Monografia* 4:1–512.

Ferrara, F., and S. Taiti. 1983. Isopodi terrestri. Contributions a l'etude de la faune granitiques de l'archipel des Seychelles. *Annales du Musee Royale Afrique Centrale, Tervuren (serie in octavo), Sciences Zoologiques* 240:1–92.

———. 1989. A new genus and species of terrestrial isopod from Malaysia (Crustacea, Oniscidea, Platyarthridae). *Journal of Natural History* 23:1033–1039.

Ferrari, F. D. 1988. Evolutionary transformations and Dollo's law. *Journal of Crustacean Biology* 8:618–619.

Ferrari, F. D., and E. L. Markhaseva. 1996. *Parkius karenwishnerae,* a new genus and species of calanoid copepod (Parkiidae, new family) from benthopelagic waters of the eastern tropical Pacific Ocean. *Proceedings of the Biological Society of Washington* 109:264–285.

Field, K. G., G. J. Olsen, D. J. Lane, S. J. Giovanni, M. T. Ghiselin, E. C. Raff, N. R. Pace, and R. A. Raff. 1988. Molecular analysis of the animal kingdom. *Science* 239:748–753.

Fiers, F. 1990. *Abscondicola humesi,* n. gen., n. sp., from the gill chambers of land crabs and the definition of Cancrincolidae n. fam. (Copepoda, Harpacticoida). *Bulletin d'Institute Royal de Sciences Naturelles Belgique (Biologie)* 60:69–103.

Forest, J. 1977. Un groupement injustifie: la superfamille des Bresilioidea. Remarques critiques sur le statut des familles sous ce nom (Crustacea Decapoda Caridea). *Bulletin de Museum National Histoire Naturelle (Paris),* series 3, no. 475, *Zoologie* 332:869–888.

———. 1987. Les Pylochelidae ou "Pagures symétriques" (Crustacea Coenobitoidea). In *Résultats des Campagnes Musorstum,* vol. 3, ed. A. Crosnier. *Mémoires du Muséum National d'Histoire Naturelle,* sére A, *Zoologie,* vol. 137, 1–254.

———. (editor). 1996. *Traité de Zoologie. Anatomie, Systématique, Biologie. Crustacés. Tome VII. Fascicule II. Généralities (suite) et systématique.* Paris: Masson, 1002 pp, 426 plates and figures, 22 tables.

———. (editor). 1999. *Traité de Zoologie. Anatomie, Systématique, Biologie. Tome VII, Fascicule IIIA. Crus-*

*tacés Péracarides.* Memoires de l'Institut Oceanographique Fondation Albert Ier, Prince de Monaco, 19:450 pp.

Forest, J., and M. de St. Laurent. 1989. Nouvelle contribution à la connaissance de *Neoglyphea inopinata* Forest & de Saint Laurent, à propos de la description de la femelle adulte. In *Résultats des Campagnes Musorstom,* vol. 5, ed. J. Forest. *Mémoires de Museum National d'Histoire Naturelle (Paris),* sère A, vol. 144, 75–92.

Fortey, R. A., D. E. G. Briggs, and M. A. Wills. 1997. The Cambrian evolutionary 'explosion' recalibrated. *BioEssays* 19:429–434.

Fortey, R. A., and R. H. Thomas (editors). 1997. *Arthropod relationships.* Systematics Association Special Volume series 55. London: Chapman and Hall, 367 pp.

Fosshagen, A., and T. M. Iliffe. 1985. Two new genera of Calanoida and a new order of Copepoda, Platycopioida, from marine caves on Bermuda. *Sarsia* 70: 345–358.

France, S. C., and T. D. Kocher. 1996. DNA sequencing of formalin-fixed crustaceans from archival research collections. *Molecular Marine Biology and Technology* 5:304–313.

Fresi, E., E. Idato, and M. B. Scipione. 1980. The Gnathostenetroidea and the evolution of primitive asellote isopods. *Monitore Zoologico Italiano (Nuovo Serie)* 14:119–136.

Frey, D. 1995. Changing attitudes toward chydorid anomopods since 1769. *Hydrobiologia* 307:43–45.

Friedrich, M., and D. Tautz. 1995. Ribosomal DNA phylogeny of the major extant arthropod classes and the evolution of myriapods. *Nature* 376:165–167.

Fryer, G. 1983. Functional ontogenetic changes in *Branchinecta ferox* (Milne-Edwards) (Crustacea: Anostraca). *Philosophical Transactions of the Royal Society of London* 303B:229–343.

———. 1985. Structure and habits of living branchiopods and their bearing on the interpretation of fossil forms. *Transactions of the Royal Society of Edinburgh* 76:103–113.

———. 1987a. Morphology and classification of the so-called Cladocera. *Hydrobiologia* 145:19–28.

———. 1987b. The feeding mechanisms of the Daphniidae (Crustacea: Cladocera): recent suggestions and neglected considerations. *Journal of Plankton Research* 9:419–432.

———. 1987c. A new classification of the branchiopod Crustacea. *Zoological Journal of the Linnean Society* 91:357–383.

———. 1987d. Quantitative and qualitative: numbers and reality in the study of living organisms. *Freshwater Biology* 17:177–189.

———. 1992. The origin of the Crustacea. *Acta Zoologica* 73:273–286.

———. 1995. Phylogeny and adaptive radiation within the Anomopoda: a preliminary exploration. *Hydrobiologia* 307:57–68.

———. 1996a. Reflections on arthropod evolution. *Biological Journal of the Linnean Society* 58:1–55.

———. 1996b. The carapace of the branchiopod Crustacea. *Philosophical Transactions of the Royal Society of London* 351B:1703–1712.

———. 1997. A defence of arthropod polyphyly. In *Arthropod relationships,* ed. R. A. Fortey and R. H. Thomas, 23–33. Systematics Association Special Volume Series 55. London: Chapman and Hall.

———. 1999a. The case of the one-eyed brine shrimp: are ancient atavisms possible? *Journal of Natural History* 33:791–798.

———. 1999b. A comment on a recent phylogenetic analysis of certain orders of the branchiopod Crustacea. *Crustaceana* 72:1039–1050.

———. 1999c. Cambrian animals: evolutionary curiosities or the crucible of creation? *Hydrobiologia* 403: 1–11.

———. 2001. The elucidation of branchiopod phylogeny. *Crustaceana* 74:105–114.

Funch, P., and R. M. Kristensen. 1995. Cycliophora is a new phylum with affinities to Entoprocta and Ectoprocta. *Nature* 378:711–714.

Galil, B. 2000. Crustacea Decapoda: review of the genera and species of the family Polychelidae Wood-Mason, 1874. In *Résultats des Campagnes Musorstom,* vol. 21, ed. A. Crosnier. *Mémoires du Muséum National d'Histoire Naturelle,* vol. 184, 285–387.

García-Machado, E., M. Pempera, N. Dennebouy, M. Oliver-Suarez, J. C. Mounolou, and M. Monnerot. 1999. Mitochondrial genes collectively suggest the paraphyly of Crustacea with respect to Insecta. *Journal of Molecular Evolution* 49:142–149.

Garey, J. R. 2000. Invertebrate evolution. In *McGraw-Hill Yearbook of Science and Technology 2001,* 216–218. New York: McGraw-Hill.

Garey, J. R., M. Krotec, D. R. Nelson, and J. Brooks. 1996. Molecular analysis supports a tardigrade-arthropod association. *Invertebrate Biology* 115:79–88.

Giribet, G., and C. Ribera. 2000. A review of arthropod phylogeny: new data based on ribosomal DNA sequences and direct character optimization. *Cladistics* 16:204–231.

Glaessner, M. F. 1969. Decapoda. In *Treatise on invertebrate paleontology, part R, Arthropoda 4,* ed. R. C. Moore, R399–R651. Lawrence, Kansas: The Geological Society of America and the University of Kansas Press.

Glenner, H., M. J. Grygier, J. T. Høeg, P. G. Jensen, and F. R. Schram. 1995. Cladistic analysis of the Cirripedia Thoracica. *Zoological Journal of the Linnean Society* 114:365–404.

Grenier, J. K., T. L. Garber, R. Warren, P. M. Whittington, and S. Carroll. 1997. Evolution of the entire arthropod *Hox* gene set predated the origin and radiation of the onychophoran/arthropod clade. *Current Biology* 7:547–553.

Griffin, D. J. G., and H. A. Tranter. 1986. The Decapoda Brachyura of the Siboga Expedition. Part VIII. Majidae. *Siboga-Expeditie Monograph 39, C4, Livraison* 148:1–335

Griffis, R. B., and T. H. Suchanek. 1991. A model of burrow architecture and trophic models in thalassinidean shrimp (Decapoda: Thalassinidea). *Marine Ecology Progress Series* 79:171–183.

Gruner, H. E. 1993. *Lehrbuch der Speziellen Zoologie begründet von Alfred Kaestner. Band. 1: Wirbellose Tiere. Teil 4: Arthropoda (ohne Insecta).* Jena: Gustav Fischer Verlag. 1209 pp.

———. 1996. Classe de Branchioures (Branchiura Thorell, 1864). In *Traité de Zoologie. Anatomie, Systématique, Biologie. Crustacés. Tome VII, Fascicule II. Généralités (suite) et Systématique,* ed. J. Forest, 739–754. Paris: Masson, 1002 pp.

Grygier, M. J. 1981. Sperm of the ascothoracican parasite *Dendrogaster,* the most primitive found in Crusta-

cea. *International Journal of Invertebrate Reproduction* 3:65–73.
———. 1982. Sperm morphology in Ascothoracida (Crustacea: Maxillopoda): confirmation of generalized nature and phylogenetic importance. *International Journal of Invertebrate Reproduction* 4:323–332.
———. 1983a. Ascothoracida and the unity of Maxillopoda. In *Crustacean issues 1. Crustacean phylogeny,* ed. F. R. Schram, 73–104. Rotterdam: A. A. Balkema Press, 372 pp.
———. 1983b. Revision of *Synagoga* (Crustacea: Maxillopoda: Ascothoracida). *Journal of Natural History* 17:213–239.
———. 1985. Comparative morphology and ontogeny of the Ascothoracida, a step toward a phylogeny of the Maxillopoda. *Dissertation Abstracts International* 45:2466B–2467B.
———. 1987a. New records, external and internal anatomy, and systematic position of Hansen's y-larvae (Crustacea: Maxillopoda: Facetotecta). *Sarsia* 72: 261–278.
———. 1987b. Classification of the Ascothoracida (Crustacea). *Proceedings of the Biological Society of Washington* 100:452–458.
———. 1987c. Nauplii, antennular ontogeny, and the position of the Ascothoracida within the Maxillopoda. *Journal of Crustacean Biology* 7:87–104.
———. 1996a. Sous-classe des Facetotecta (Facetotecta Grygier, 1985). In *Traité de Zoologie. Anatomie, Systématique, Biologie. Crustacés. Tome VII, Fascicule II. Généralités (suite) et Systématique,* ed. J. Forest, 425–432. Paris: Masson, 1002 pp.
———. 1996b. Sous-classe de Ascothoracides (Ascothoracida Lacaze-Duthiers, 1880). In *Traité de Zoologie. Anatomie, Systématique, Biologie. Crustacés. Tome VII, Fascicule II. Généralités (suite) et Systématique,* ed. J. Forest, 433–452. Paris: Masson, 1002 pp.
Grygier, M. J., and T. E. Bowman. 1990. The correct family-level name for the "cryptoniscid" isopods (Epicaridea). *Crustaceana* 58:27–32.
———. 1991. The authorship of Cryptoniscidae (Isopoda, Epicaridea); a correction. *Crustaceana* 61:106–107.
Guinot, D. 1977. Propositions pour une nouvelle classification des Crustacés Décapodes Brachyoures. *Comptes Rendus Hebdomadaires des Seances de l'Academie des Sciences,* series 3, 285:1049–1052.
———. 1978. Principes d'une classification évolutive des Crustacés Décapodes Brachyoures. *Bulletin Biologique de le France et de le Belgique (n. s.)* 112:211–292.
———. 1979. Données nouvelles sur la morphologie, la phylogenèse et la taxonomie des Crustacés Décapodes Brachyoures. *Mémoires du Muséum National d'Histoire Naturelle (Paris)* sére A, *Zoologie* 112:1–354.
———. 1985. Crustacea. In *French Polynesian coral reefs, reef knowledge and field guides, 5th International Coral Reef Congress, Tahiti,* ed. G. Richard, 446–455. Paris: Muséum National d'Histoire Naturelle and École Pratique des Hautes Études Antenne de Tahiti.
———. 1988. Les crabes des sources hydrothermales de la dorsale du Pacifique oriental (Campagne BIOCYARISE), 1984. *Oceanologica Acta* (special volume) 8:109–118.
———. 1990. *Austinograea alayseae* sp. nov., crabe hydrothermal découvert dans le bassin de Lau, Pacifique sud-occidental (Crustacea Decapoda Brachyura). *Bulletin du Museum National d'Histoire (Paris)* 11:879–903.
———. 1991. Établissement de la famille des Poupiniidae pour *Poupinia hirsuta* gen. nov., sp. nov. de Polynésie (Crustacea Decapoda Brachyura Homoloidea). *Bulletin du Museum National d'Historie Naturelle (Paris)* 12:577–605.
———. 1993. Donnees nouvelles sur les Raninoidea de Haan, 1841 (Crustacea Decapoda Brachyura Podotremata). *Comptes Rendus Hebdomadaires des Seances de l'Academie des Sciences,* series 3, 316: 1324–1331.
———. 1995. Crustacea Decapoda Brachyura: révision des Homolodromiidae Alcock, 1900. In *Résultats des Campagnes Musorstom,* vol. 13, ed. A. Crosnier. *Mémoires du Muséum National d'Histoire Naturelle,* vol. 163, 155–282.
Guinot, D., and J.-M. Bouchard. 1998. Evolution of the abdominal holding systems of brachyuran crabs (Crustacea, Decapoda, Brachyura). *Zoosystema* 20: 613–694.
Guinot, D., B. G. M. Jamieson, and B. Richer de Forges. 1994. Relationship of Homolidae and Dromiidae—evidence from spermatozoal ultrastructure (Crustacea, Decapoda). *Acta Zoologica* 75:255–267.
Guinot, D., B. G. M. Jamieson, B. Richer de Forges, and C. C. Tudge. 1998. Comparative spermatozoal ultrastructure of the three dromiacean families exemplified by *Homolodromia kai* (Homolodromiidae), *Sphaerodromia lamellata* (Dromiidae), and *Dynomene tanensis* (Dynomenidae) (Podotremata: Brachyura). *Journal of Crustacean Biology* 18:78–94.
Guinot, D., B. G. M. Jamieson, and C. C. Tudge. 1997. Ultrastructure and relationships of spermatozoa of the freshwater crabs *Potamon fluviatile* and *Potamon ibericum* (Crustacea, Decapoda, Potamidae). *Journal of Zoology* 241:229–244.
Guinot, D., and B. Richer de Forges. 1995. Crustacea Decapoda Brachyura: révision des Homolidae de Haan, 1839. In *Résultats des Campagnes Musorstom,* vol. 13, ed. A. Cronsnier. *Mémoires de Muséum National d'Histoire Naturelle,* vol. 163, 283–517.
———. 1997. Affinités entre les Hymenosomatidae MacLeay, 1838 et les Inachoididae Dana, 1851 (Crustacea, Decapoda, Brachyura). *Zoosystema* 19:453–502.
Gutu, M. 1996. Tanaidacea (Crustacea, Peracarida) from Brazil, with description of new taxa and systematical remarks on some families. *Travaux du Muséum National d'Histoire Naturelle "Grigore Antipa"* 36:23–133.
———. 1998. Spelaeogriphacea and Mictacea (partim) suborders of a new order, Cosinzeneacea (Crustacea, Peracarida). *Travaux du Muséum National d'Histoire Naturelle "Grigore Antipa"* 40:121–129.
Gutu, M., and T. M. Iliffe. 1998. Description of a new hirsutiid (n. g., n. sp.) and reassignment of this family from Order Mictacea to the new order, Bochusacea (Crustacea, Peracarida). *Travaux du Muséum d'Histoire Naturelle "Grigore Antipa"* 40:93–120.
Gutu, M., and J. Sieg. 1999. Ordre des Tanaïdacés (Tanaidacea Hansen, 1895). In *Traité de Zoologie. Anatomie, Systématique, Biologie. Tome VII, Fascicule IIIA. Crustacés Péracarides,* ed. J. Forest. Memoires de l'Institut Oceanographique Fondation Albert Ier, Prince de Monaco, 19:353–389.
Hanner, R., and M. Fugate. 1997. Branchiopod phyloge-

netic reconstruction from 12S rDNA sequence data. *Journal of Crustacean Biology* 17:174–183.

Hansen, H. J. 1895. Isopoden, Cumaceen und Stomatopoden der Plankton Expedition. *Ergebnisse der im Atlanishcen Ozean Plankton-Expedition der Humbolt-Stiftung, 1889* 2:1–105.

Harling, C. 2000. Reexamination of eye design in the classification of stomatopod crustaceans. *Journal of Crustacean Biology* 20:172–185.

Harris, D. J., L. S. Maxson, L. F. Braithwaite, and K. A. Crandall. 2000. Phylogeny of the thoracican barnacles based on 18S rDNA sequences. *Journal of Crustacean Biology* 20:393–398. Harrison, K., and J. P. Ellis. 1991. The genera of the Sphaeromatidae (Crustacea: Isopoda): a key and distribution list. *Invertebrate Taxonomy* 5:915–952.

Hart, C. W. J., and R. B. Manning. 1986. Two new shrimps (Procarididae and Agostocarididae, new family) from marine caves of the western North Atlantic. *Journal of Crustacean Biology* 6:408–416.

Hartmann, G., and M.-C. Guillaume. 1996. Classe des Ostracodes (Ostracoda Latreille, 1802). In *Traité de Zoologie. Anatomie, Systématique, Biologie. Crustacés. Tome VII, Fascicule II. Généralités (suite) et Systématique,* ed. J. Forest, 755–839. Paris: Masson, 1002 pp.

Harvey, A. H., J. W. Martin, and R. Wetzer. In press. Crustacea. In *Atlas of marine invertebrate larvae,* ed. C. Young, M. Sewell, and M. Rice. London: Academic Press.

Harzsch, S., and D. Walossek. 2001. Neurogenesis in the developing visual system of the branchiopod crustacean *Triops longicaudatus* (LeConte, 1846): corresponding patterns of compound-eye formation in Crustacea and Insecta? *Developmental Genes and Evolution* 211:37–43.

Hay, W. P., and C. A. Shore. 1918. The decapod crustaceans of Beaufort, N. C., and the surrounding regions. *Bulletin of the U.S. Fisheries Bureau* 35(for 1915 and 1916):369–475.

Haye, P., and I. Kornfield. 1999. Familial relationships within the Cumacea: preliminary molecular findings. *Program and Abstracts, The Crustacean Society 1999 Summer Meeting, Lafayette, Louisiana*: 33.

Hendrickx, M. E. 1995. Checklist of brachyuran crabs (Crustacea: Decapoda) from the eastern tropical Pacific. *Bulletin de l'Institut Royal des Sciences Naturelles de Belgique, Biologie* 65:125–150.

———. 1998. A new genus and species of "goneplacid-like" brachyuran crab (Crustacea: Decapoda) from the Gulf of California, Mexico, and a proposal for the use of the family Pseudorhombilidae Alcock, 1900. *Proceedings of the Biological Society of Washington* 111:634–644.

———. 1999. *Los Cangrejos Braquiros (Crustacea: Brachyura: Majoidea y Parthenopoidea) del Pacifico Mexicano.* Instituto de Ciencias del Mar y Limnología, Universidad Nacional Autónoma de México: CONABIO, 274 pp., 13 plates.

Hessler, R. R. 1982. The structural morphology of walking mechanisms in eumalacostracan crustaceans. *Philosophical Transactions of the Royal Society of London* 296B:245–298.

———. 1983. A defense of the caridoid facies; wherein the early evolution of the Eumalacostraca is discussed. In *Crustacean issues 1. Crustacean phylogeny,* ed. F. R. Schram, 145–164. Rotterdam: A. A. Balkema Press, 372 pp.

———. 1984. *Dahlella caldariensis,* new genus, new species: a leptostracan (Crustacea, Malacostraca) from deep-sea hydrothermal vents. *Journal of Crustacean Biology* 4:655–664.

———. 1985. Swimming in Crustacea. *Transactions of the Royal Society of Edinburgh* 76:115–122.

———. 1992. Reflections on the phylogenetic position of the Cephalocarida. *Acta Zoologica* 73:315–316.

———. 1999. Ordre des Mictacés (Mictacea Bowman, Garner, Hessler, Iliffe, Sanders, 1985). In *Traité de Zoologie. Anatomie, Systématique, Biologie. Tome VII, Fascicule IIIA. Crustacés Péracarides,* ed. J. Forest. Memoires de l'Institut Oceanographique Fondation Albert Ier, Prince de Monaco, 19:87–91.

Hessler, R. R., and R. Elofsson. 1996. Classe de Céphalocarides (Cephalocarida Sanders, 1955). In *Traité de Zoologie. Anatomie, Systematique, Biologie. Tome VII, Fascicule II. Généralités (suite) et Systématique,* ed. J. Forest, 271–286. Paris: Masson, 1002 pp.

Hessler, R. R., and J. W. Martin. 1989. *Austinograea williamsi,* new genus, new species, a hydrothermal vent crab (Decapoda: Bythograeidae) from the Mariana Back-Arc Basin, western Pacific. *Journal of Crustacean Biology* 9:645–661.

Hessler, R. R., and L. Watling. 1999. Les Péracarides: un groupe controversé. In *Traité de Zoologie. Anatomie, Systématique, Biologie. Tome VII, Fascicule IIIA. Crustacés Péracarides,* ed. J. Forest. Memoires de l'Institut Oceanographique Fondation Albert Ier, Prince de Monaco, 19:1–10.

Heymons, R. 1935. Pentastomida. In *Bronns Klassen und Ordnungen das Tierreichs,* vol. 5, 1–268. Leipzig: Akademische Verlagsgesellschaft.

Hibbett, D. S., and M. J. Donoghue. 1998. Integrating phylogenetic analysis and classification in fungi. *Mycologia* 90:347–356.

Hickman, C. P., Jr., L. S. Roberts, and A. Larson. 1996. *Integrated principles of zoology.* Dubuque, Iowa: Wm. C. Brown Publishers, Times Mirror Higher Education Group, 901 pp.

Ho, J.-S. 1984. New family of poecilostomatoid copepods (Spiophanicolidae) parasitic on polychaetes from southern California, with a phylogenetic analysis of nereicoliform families. *Journal of Crustacean Biology* 4:134–146.

———. 1990. Phylogenetic analysis of copepod orders. *Journal of Crustacean Biology* 10:528–536.

———. 1991. Phylogeny of Poecilostomatoida: a major order of symbiotic copepods. Proceedings of the 4th International Conference on Copepoda. *Bulletin of the Plankton Society of Japan,* Special Volume, 25–48.

———. 1994a. Copepod phylogeny: a reconsideration of Huys & Boxshall's "parsimony versus homology." *Hydrobiologia* 292/293:31–39.

———. 1994b. Origin and evolution of the parasitic cyclopoid copepods. *International Journal for Parasitology* 24:1293–1300.

Ho, J.-S., M. Conradi, and P. J. López-González. 1998. A new family of cyclopoid copepods (Fratiidae) symbiotic in the ascidian (*Clavellina dellavallei*) from Cádiz, Spain. *Journal of Zoology* 246:39–48.

Ho, J.-S., and I. H. Kim. 1997. A new family of poecilostomatoid copepods (Polyankyliidae) from a tide pool on a mud flat in Korea. *Korean Journal of Biological Sciences* 1:429–434.

Ho, J.-S., and V. E. Thatcher. 1989. A new family of cyclopoid copepods (Ozmanidae) parasitic in the hemocoel of a snail from the Brazilian Amazon. *Journal of Natural History* 23:903–911.

Høeg, J. T. 1992a. The phylogenetic position of the Rhizocephala: are they truly barnacles? *Acta Zoologica* 73:271–392.

———. 1992b. Rhizocephala. In *Microscopic anatomy of invertebrates*. vol. 9, *Crustacea*, ed F. W. Harrison and A. G. Humes, 313–345. New York; Wiley-Liss.

———. 1995. Sex and the single cirriped: a phylogenetic perspective. In *Crustacean issues 10. New frontiers in barnacle evolution*, ed. F. R. Schram and J. T. Høeg, 195–207. Rotterdam: A. A. Balkema Press, 318 pp.

Høeg, J. T., B. Hosfeld, and P. G. Jensen. 1998. TEM studies on the lattice organs of cirripede cypris larvae (Crustacea, Thecostraca, Cirripedia). *Zoomorphology* 118:195–205.

Høeg, J. T., and J. Lützen. 1993. Comparative morphology and phylogeny of the family Thompsoniidae (Cirripedia, Rhizocephala, Akentrogonida), with descriptions of three new genera and seven new species. *Zoologica Scripta* 22:363–386.

———. 1996. Super-ordre des Rhizocephales (Rhizocephala F. Müller, 1862). In *Traité de Zoologie. Anatomie, Systématique, Biologie. Crustacés. Tome VII, Fascicule II. Généralités (suite) et Systématique*, ed. J. Forest, 541–570. Paris: Masson, 1002 pp.

Høeg, J. T., and A. V. Rybakov. 1992. Revision of the Rhizocephala Akentrogonida (Cirripedia), with a list of all the species and a key to the identification of families. *Journal of Crustacean Biology* 12:600–609.

Høeg, J. T., M. A. Whyte, H. Glenner, and F. R. Schram. 1999. New evidence on the basic phylogeny of the Cirripedia Thoracia. In *Crustaceans and the biodiversity crisis. Proceedings of the 4th International Crustacean Congress, Amsterdam*, ed. F. R. Schram and J. C. von Vaupel Klein, 101–149. Leiden: Brill.

Hof, C. H. J. 1998a. Mesozoic stomatopods. *Proceedings and Abstracts of the 4th International Crustacean Congress, Amsterdam*: 78 (abstract 123).

———. 1998b. Fossil stomatopods (Crustacea: Malacostraca) and their phylogenetic impact. *Journal of Natural History* 32:1567–1576.

Hof, C. H. J., and F. R. Schram. 1999. The phylogenetic position and evolution of Hoplocarida. *Program and Abstracts, The Crustacean Society 1999 Summer Meeting, Lafayette, Louisiana*: 35.

Hogans, W. E., and G. W. Benz. 1990. A new family of parasitic copepods, the Lernaeosoleidae (Poecilostomatoida), from demersal fishes in the northwest Atlantic, with a description of *Bobkabata kabatabobbus* n. gen., n. sp. and a redescription of *Lernaeosolea lycodis* Wilson, 1944. *Canadian Journal of Zoology* 68:2483–2488.

Holdich, D., R. Lincoln, and J. Ellis. 1984. The biology of terrestrial isopods: terminology and classification. *Symposium of the Zoological Society of London* 53: 1–6.

Holsinger, J. R. 1989. Allocrangonyctidae and Pseudocrangonyctidae, two new families of Holarctic subterranean amphipod crustaceans (Gammaridea), with comments on their phylogenetic and zoogeographic relationships. *Proceedings of the Biological Society of Washington* 102:947–959.

———. 1992. Sternophysingidae, a new family of subterranean amphipods (Gammaridea: Crangonyctoidea) from South Africa, with descriptions of *Sternophysinx calceola*, new species, and comments on phylogenetic and biogeographic relationships. *Journal of Crustacean Biology* 12:111–124.

Holthuis, L. B. 1955. The Recent genera of the caridean and stenopodidean shrimps (Class Crustacea, Order Decapoda, supersection Natantia) with keys for their determination. *Zoologische Verhandelingen* 26:1–157.

———. 1991. *Marine lobsters of the world. An annotated and illustrated catalogue of species of interest to fisheries known to date*. FAO species catalogue, vol. 13, and FAO Fisheries Synopsis no. 125. Rome: FAO, i–vii + 292 pp.

———. 1993a. *The Recent genera of the caridean and stenopodidean shrimps (Crustacea, Decapoda) with an appendix on the order Amphionidacea*. Leiden: Nationaal Natuurhistorisch Museum, 328 pp.

———. 1993b. The non-Japanese new species established by W. de Haan in the Crustacea volume of Fauna Japonica (1833–1850). In *Ph. F. von Siebold and Natural History of Japan, Crustacea*, ed. T. Yamaguchi, 599–642. Tokyo: The Carcinological Society of Japan.

Hou X., and J. Bergström. 1991. The arthropods of the Lower Cambrian Chengjiang fauna, with relationships and evolutionary significance. In *The early evolution of Metazoa and the significance of problematic taxa*, ed. A. M. Simonetta and S. Conway Morris, 179–188. London: Cambridge University Press.

———. 1997. Arthropods of the Lower Cambrian Chengjiang fauna of southwest China. *Fossils and Strata* 45:1–116.

Humes, A. G. 1986. *Myicola metisiensis* (Copepoda: Poecilostomatoida), a parasite of the bivalve *Mya arenaria* in eastern Canada, redefinition of the Myicolidae, and diagnosis of the Anthessiidae n. fam. *Canadian Journal of Zoology* 64:1021–1033.

———. 1987. Copepoda from deep-sea hydrothermal vents. *Bulletin of Marine Science* 41:645–788.

Humes, A. G., and G. A. Boxshall. 1996. A revision of the lichomolgoid complex (Copepoda: Poecilostomatoidea), with the recognition of six new families. *Journal of Natural History* 30:175–227.

Humes, A. G., and J. H. Stock. 1991. Coralliomyzontidae, fam. n. (Copepoda: Siphonostomatoida), associated with scleractinian corals in Madagascar. *Bulletin Zoölogisch Museum (Universiteit van Amsterdam)* 13:17–24.

Huvard, A. 1990. Ultrastructural study of the naupliar eye of the ostracode *Vargula graminicola* (Crustacea, Ostracoda). *Zoomorphology* 110:47–51.

Huys, R. 1988a. Gelyelloida, a new order of stygobiont copepods from European karstic systems. *Hydrobiologia* 167/168:485–495.

———. 1988b. On the identity of the Namakosiramiidae Ho & Perkins, 1977 (Crustacea, Copepoda), including a review of harpacticoid associates of Echinodermata. *Journal of Natural History* 22:1517–1532.

———. 1990a. *Adenopleurella*, new genus, *Proceropes*, new genus, *Sarsocletodes* Wilson (ex Laophontidae), and *Miroslavia* Apostolov (ex Cletodidae): representatives of a new family (Copepoda: Harpacticoida). *Journal of Crustacean Biology* 10:340–363.

———. 1990b. A new harpacticoid copepod family collected from Australian sponges and the status of the subfamily Rhynchothalestrinae Lang. *Zoological Journal of the Linnean Society* 99:51–115.

———. 1990c. Amsterdam expeditions to the West Indian Islands, Report 64. A new family of harpacticoid copepods and an analysis of the phylogenetic relationships within the Laophontoidea T. Scott. *Bijdragen tot de Dierkunde* 60:79–120.

———. 1990d. Allocation of the Mantridae Leigh-Sharpe to the Cyclopoida (Crustacea: Copepoda) with notes on *Nearchinotodelphys* Ummerkutty. *Bijdragen tot de Dierkunde* 60:283–291.

———. 1990e. *Campyloxiphos dineti* gen. et spec. nov. from off Namibia and a redefinition of the Deoterthridae Boxshall and Lincoln (Crustacea: Tantulocarida). *Journal of Natural History* 24:415–432.

———. 1991. Tantulocarida (Crustacea: Maxillopoda): a new taxon from the temporary meiobenthos. *Marine Ecology (Publicazioni della Stazione Zoologica di Napoli I)* 12:1–34.

———. 1992. The amphiatlantic distribution of *Leptastacus macronyx* (T. Scott, 1892) (Copepoda: Harpacticoida): a paradigm of taxonomic confusion; and, a cladistic approach to the classification of the Leptastacidae Land, 1948. *Mededelingen Koninklijke Academie voor Wetenschappen, Letteren en Schone Kunsten van Belgie* 54:21–196.

———. 1993. Styracothoracidae (Copepoda: Harpacticoida), a new family from the Philippine deep sea. *Journal of Crustacean Biology* 13:769–783.

———. 1997. Superornatiremidae fam. nov. (Copepoda: Harpacticoida): an enigmatic family from North Atlantic anchialine caves. *Scientia Marina* 60:497–542.

———. 2001. Splanchnotrophid systematics: a case of polyphyly and taxonomic myopia. *Journal of Crustacean Biology* 21:106–156.

Huys, R., and R. Böttger-Schnack. 1997. On the diphyletic origin of the Oncaeidae Giesbrech, 1892 (Copepoda: Poecilostomatoida) with a phylogenetic analysis of the Lubockiidae fam. nov. *Zoologischer Anzeiger* 235:243–261.

Huys, R., and G. A. Boxshall. 1991. *Copepod evolution*. London: The Ray Society, 468 pp.

Huys, R., G. A. Boxshall, and R. J. Lincoln. 1993. The tantulocarid life cycle: the circle closed? *Journal of Crustacean Biology* 13:432–442.

Huys, R., and J. M. Gee. 1996. A revision of *Danielssenia* Boeck and *Psammis* Sars with the establishment of two new genera *Archisenia* and *Bathypsammis* (Harpacticoida: Paranannopidae). *Bulletin National d'Histoire Naturelle (Zoologie)* 59:45–81.

Huys, R., J. M. Gee, C. G. Moore, and R. Hammond. 1996. *Marine and brackish water harpacticoid copepods. Part 1. Keys and notes for identification of the species. Synopses of the British Fauna (New Series)*. London: Academic Press, for the Linnean Society of London, 352 pp. Huys, R., and T. M. Iliffe. 1998. Novocriniidae, a new family of harpacticoid copepods from anchialine caves in Belize. *Zoologica Scripta* 27:1–15.

Huys, R., and W. Lee. 1999. On the relationships of the Normanellidae and the recognition of Cletopsyllidae grad. nov. (Copepoda, Harpacticoida). *Zoologischer Anzeiger* 237:267–290.

Huys, R., S. Ohtsuka, and G. A. Boxshall. 1994. A new tantulocaridan (Crustacea: Maxillopoda) parasitic on calanoid, harpacticoid, and cyclopoid copepods. *Publications of the Seto Marine Biological Laboratory* 36:197–209.

Huys, R., and K. A. Willems. 1989. *Laophontopsis* Sars and the taxonomic concept of the Normanellinae (Copepoda: Harpacticoida): a revision. *Bijdragen tot de Dierkunde* 59:203–227.

Illg, P. L., and P. L. Dudley. 1980. The family Ascidicolidae and its subfamilies (Copepoda, Cyclopoida), with descriptions of new species. *Memoires du Museum National d'Histoire Naturelle, Nouvelle Serie A, Zoologie* 117:1–192.

International Commission on Zoological Nomenclature. 1985a. *International Code of Zoological Nomenclature*, 3rd edition. London: The International Trust for Zoological Nomenclature and The British Museum (Natural History), 338 pp.

———. 1985b. Opinion 1357. Anuropodidae Băcescu, 1980 (Crustacea, Tanaidacea) and Anuropodidae Stebbing, 1893 (Crustacea, Isopoda); a ruling to remove the homonymy. *Bulletin of Zoological Nomenclature* 42:338–340.

———. 1999. *International Code of Zoological Nomenclature*, 4th edition. London: The International Trust for Zoological Nomenclature and The Natural History Museum, 306 pp.

*International Code of Zoological Nomenclature*, 3rd edition, adopted by the XX General Assembly of the International Union of Biological Sciences. 1985. Berkeley, California: University of California Press and International Trust for Zoological Nomenclature, 338 pp.

Ishimaru, S. 1994. A catalogue of gammaridean and ingolfiellidean Amphipoda recorded from the vicinity of Japan. *Reports of the Seto Marine Biological Laboratory* 24:29–846.

Itô, T. 1989. Origin of the basis in copepod limbs, with reference to remipedian and cephalocarid limbs. *Journal of Crustacean Biology* 9:85–103.

Itô, T., and F. R. Schram. 1988. Gonopores and the reproductive system of nectiopodan Remipedia. *Journal of Crustacean Biology* 8:250–253.

Ivanenko, V. N., F. D. Ferrari, and A. V. Smurov. 2001. Nauplii and copepodids of *Scottomyzon gibberum* (Copepoda: Siphonostomatoida: Scottomyzontidae, a new family), a symbiont of *Asterias rubens* (Asteroida). *Proceedings of the Biological Society of Washington* 114:237–261.

Iverson, E. 1982. Revision of the isopod family Sphaeromatidae (Crustacea: Isopoda: Flabellifera) I. Subfamily names with diagnoses and key. *Journal of Crustacean Biology* 2:248–254.

Izawa, K. 1996. *Archidactylina myxinicola*, new genus, new species (Siphonostomatoida), in a new family of Copepoda parasitic on hagfishes (Agnatha: Myxiniformes) from Japan. *Journal of Crustacean Biology* 16:406–417.

Jamieson, B. G. M. 1987. *The ultrastructure and phylogeny of insect spermatozoa*. London: Cambridge University Press, 320 pp.

———. 1989a. The ultrastructure of the spermatozoa of four species of xanthid crabs (Crustacea, Brachyura, Xanthidae). *Journal of Submicroscopic Cytology and Pathology* 21:589–584.

———. 1989b. Ultrastructural comparison of the spermatozoa of *Ranina ranina* (Oxystomata) and of other crabs exemplified by *Portunus pelagicus* (Brachygnatha) (Crustacea, Brachyura). *Zoomorphology* 109:103–111.

———. 1989c. A comparison of the spermatozoa of *Oratosquilla stephensoni* and *Squilla mantis* (Crustacea, Stomatopoda) with comments on the phylogeny of the Malacostraca. *Zoologica Scripta* 18:509–517.

———. 1990. The ultrastructure of the spermatozoa of *Petalomera lateralis* (Gray) (Crustacea, Brachyura,

Dromiacea) and its phylogenetic significance. *Invertebrate Reproduction and Development* 17:39–45.

———. 1991a. Ultrastructure and phylogeny of crustacean spermatozoa. In *Proceedings of the 1990 International Crustacean Conference,* ed. P. J. F. Davie and R. H. Quinn. *Memoirs of the Queensland Museum* 31:109–142.

———. 1991b. Sperm and phylogeny in the Brachyura (Crustacea). In *Comparative spermatology 20 years after,* ed. B. Bacetti, 967–972. New York: Serono Symposia Publications from Raven Press, vol. 75.

———. 1993. Spermatological evidence for the taxonomic status of *Trapezia* (Crustacea: Brachyura: Heterotremata). *Memoirs of the Queensland Museum* 33: 225–234.

———. 1994. Phylogeny of the Brachyura with particular reference to the Podotremata: evidence from a review of spermatozoal ultrastructure. *Philosophical Transactions of the Royal Society* 345B:373–393.

Jamieson, B. G. M., D. Guinot, and B. Richer de Forges. 1993a. The ultrastructure of the spermatozoon of *Paradynomene tuberculata* Sakai, 1963 (Crustacea, Brachyura, Dynomenidae): synapomorphies with dromiid sperm. *Helgoländer Meeresuntersuchungen* 47:311–322.

———. 1993b. Spermatozoal ultrastructure in 4 genera of Homolidae (Crustacea, Decapoda)—exemplified by *Homologenus* sp., *Latreillopsis* sp., *Homolomannia sibogae,* and *Paromolopsis boasi. Helgoländer Meeresuntersuchungen* 47:323–334.

———. 1993c. The spermatozoon of *Calocarcinus africanus* (Heterotremata, Brachyura, Crustacea)—ultrastructural synapomorphies with xanthid sperm. *Invertebrate Reproduction and Development* 24: 189–196.

———. 1994a. Relationships of the Cyclodorippoidea Ortmann—evidence from spermatozoal ultrastructure in the genera *Xeinostoma, Tymolus,* and *Cymonomus* (Crustacea, Decapoda). *Invertebrate Reproduction and Development* 26:153–164.

———. 1994b. Podotreme affinities of *Raninoides* sp. and *Lyreidus brevifrons*—evidence from spermatozoal ultrastructure (Crustacea, Brachyura, Raninoidea). *Marine Biology* 120:239–249.

———. 1995. Phylogeny of the Brachyura: evidence from spermatozoal ultrastructure (Crustacea, Decapoda). In *Advances in spermatozoal phylogeny and taxonomy,* ed. B. G. M. Jamieson, J. Ausio, and J.-L. Justine *Mémoires du Muséum National d'Histoire Naturelle,* vol. 166, 265–283.

———. 1996. Contrasting spermatozoal ultrastructure in two thoracotreme crabs, *Cardisoma carnifex* (Gecarcinidae) and *Varuna litterata* (Grapsidae) (Crustacea: Brachyura). *Invertebrate Reproduction and Development* 29:111–126.

Jamieson, B. G. M., D. Guinot, C. C. Tudge, and B. Richer de Forges. 1997. Ultrastructure of the spermatozoa of *Corystes cassivelaunus* (Corystidae), *Platepistoma nanum* (Cancridae), and *Cancer pagurus* (Cancridae) supports recognition of the Corystoidea (Crustacea, Brachyura, Heterotremata). *Helgoländer Meeresuntersuchungen* 51:83–93.

Jamieson, B. G. M., and C. C. Tudge. 1990. Dorippids are Heterotremata: evidence from ultrastructure of the spermatozoa of *Neodorippe astuta* (Dorippidae) and *Portunus pelagicus* (Portunidae) (Brachyura: Decapoda). *Marine Biology* 106:347–354.

Jamieson, B. G. M., C. C. Tudge, and D. M. Scheltinga. 1993. The ultrastructure of the spermatozoan of *Dromidiopsis edwardsi* Rathbun, 1919 (Crustacea: Brachyura: Dromiidae): confirmation of a dromiid sperm type. *Australian Journal of Zoology* 41:537–548.

Jarett, N. E., and E. L. Bousfield. 1994a. The amphipod superfamily Phoxocephaloidea on the Pacific Coast of North America. Family Phoxocephalidae. Part I. Metharpiniinae, new subfamily. *Amphipacifica* 1: 58–140.

———. 1994b. The amphipod superfamily Phoxocephaloidea on the Pacific Coast of North America. Family Phoxocephalidae. Part II. Subfamilies Pontharpiniinae, Parharpiniinae, Brolginae, Phoxocephalinae, and Harpiniinae. Systematics and distributional ecology. *Amphipacifica* 1:71–150.

Jarman, S. N., S. Nicol, N. G. Elliott, and A. McMinn. 2000. 28S rDNA evolution in the Eumalacostraca and the phylogenetic position of krill. *Molecular Phylogenetics and Evolution* 17:26–36.

Jenner, R. A., C. H. J. Hof, and F. R. Schram. 1998. Palaeo- and archaeostomatopods (Hoplocarida, Crustacea) from the Bear Gulch Limestone, Mississippian (Namurain), of central Montana. *Contributions to Zoology* (Amsterdam: SPB Academic Publishing) 67: 155–185.

Jensen, P. G., J. T. Høeg, S. Bower, and A. V. Rybakov. 1994a. Scanning electron microscopy of lattice organs in cyprids of the Rhizocephala Akentrogonida (Crustacea: Cirripedia). *Canadian Journal of Zoology* 72:1018–1026.

Jensen, P. G., J. Moyse, J. T. Høeg, and H. Al-Yahya. 1994b. Comparative SEM studies of lattice organs: putative sensory structures on the carapace of larvae from Ascothoracida and Cirripedia (Crustacea Maxillopoda Theocostraca). *Acta Zoologica* 75:125–142.

Jespersen, Å. 1979. Spermiogenesis in two species of *Nebalia* Leach (Crustacea, Malacostraca, Phyllocarida). *Zoomorphology* 93:87–97.

Jones, M. L. 1961. *Lightiella serendipita* gen. nov., sp. nov., a cephalocarid from San Francisco Bay, California. *Crustaceana* 3:31–46.

Jones, N. S. 1976. *British Cumaceans. Synopses of the British Fauna No. 7.* London: Academic Press, for the Linnean Society of London, 66 pp.

Just, J. 1990. Abyssianiridae, a synonym of Paramunnidae (Crustacea: Isopoda: Asellota) with two new species of *Abyssianira* from south-eastern Australia. *Memoirs of the Museum of Victoria* 50:403–416.

Just, J., and G. C. B. Poore. 1988. Second record of Hirsutiidae (Peracarida: Mictacea): *Hirsutia sandersetalia,* new species, from southeastern Australia. *Journal of Crustacean Biology* 8:483–488.

———. 1992. Vermectiadidae, a new primitive asellote isopod family with important phylogenetic implications. *Journal of Crustacean Biology* 12:125–144.

Kamaltynov, R. M. 1999. On the higher classification of Lake Baikal amphipods. *Crustaceana* 72:933–944.

Kensley, B. 1998. Estimates of species diversity of free-living isopod crustaceans on coral reefs. *Coral Reefs* 17:83–88.

Kensley, B., M. Schotte, and S. Schilling. 1996–1998. *World list of marine, freshwater and terrestrial isopod crustaceans.* URL gopher://nmnhgoph.si.edu:70/11/.invertebrate/.crustaceans; URL http://www.nmnh.si.edu/iz/isopod/about.html#org.

Kim, C. B., and W. Kim. 1993. Phylogenetic relationships

among gammaridean families and amphipod suborders. *Journal of Natural History* 27:933–946.

Kim, W., and L. G. Abele. 1990. Molecular phylogeny of selected decapod crustaceans based on 18S rRNA nucleotide sequences. *Journal of Crustacean Biology* 10:1–13.

Kitaura, J., K. Wada, and M. Nishida. 1998. Molecular phylogeny and evolution of unique mud-using territorial behavior in ocypodid crabs (Crustacea: Brachyura: Ocypodidae). *Molecular Biology and Evolution* 15:626–637.

Koltzoff, N. K. 1906. Studien über die Gestalt der Zelle. I. Untersuchungen über dies Spermien der Decapoden, als Einleitung in das Problem der Zellengestalt. *Archiv für Mikroskopische Anatomomie* 67:364–572.

Korn, O. M. 1995. Naupliar evidence for cirripede taxonomy and phylogeny. In *Crustacean issues 10. New frontiers in barnacle evolution,* ed. F. R. Schram and J. T. Høeg, 87–121. Rotterdam: A. A. Balkema Press, 318 pp.

Kornicker, L. S. 1986. Sarsiellidae of the western Atlantic and northern Gulf of Mexico, and revision of the Sarsiellinae (Ostracoda: Myodocopina). *Smithsonian Contributions to Zoology* 415:1–217.

Kornicker, L. S., and I. G. Sohn. 1976. Phylogeny, ontogeny, and morphology of living and fossil Thaumatocypridacea (Myodocopa: Ostracoda). *Smithsonian Contributions to Zoology* 219:1–124.

Kropp, R. K., and R. B. Manning. 1985. Cryptochiridae, the correct name for the family containing the gall crabs (Crustacea: Decapoda: Brachyura). *Proceedings of the Biological Society of Washington* 98:954–955.

———. 1987. The Atlantic gall crabs, family Cryptochiridae (Crustacea: Decapoda: Brachyura). *Smithsonian Contributions to Zoology* 462:1–21.

Kussakin, O. G. 1988. Marine and brackish isopods (Isopoda) of cold and temperate waters of the northern hemisphere. Volume 3. Suborder Asellota. Part 1. Families Janiridae, Santiidae, Dendrotionidae, Munnidae, Paramunnidae, Haplomunnidae, Mesosignidae, Halponiscidae, Mictosomatidae, Ischnomesidae. *Opredeliteli Faune SSR* 152:1–500.

Lake, J. A. 1990. Origin of the Metazoa. *Proceedings of the National Academy of Sciences of the United States of America* 87:763–766.

Lamb, E. J., G. A. Boxshall, P. J. Mill, and J. Grahame. 1996. Nucellicolidae: a new family of endoparasitic copepods (Poecilostomatoida) from the dog whelk *Nucella lapillus* (Gastropoda). *Journal of Crustacean Biology* 16:142–148.

Land, M. F. 1996. Les yeux: structure et fonctionnement des mécanismes optiques. In *Traité de Zoologie. Anatomie, Systématique, Biologie. Crustacés. Tome VII, Fascicule II. Généralités (suite) et Systématique,* ed. J. Forest, 1–42. Paris: Masson, 1002 pp.

Lange, S., F. R. Schram, C. H. J. Hof, and F. A. Steeman. In press. A new genus and species of Thylacocephala, with observations that allow assignment of the group at last to the Crustacea. *Palaeontology.*

Larsen, K., and G. D. F. Wilson. 1998. Tanaidomorphan systematics—is it obsolete? *Journal of Crustacean Biology* 18:346–362.

Latreille, P. A. 1802. *Histoire naturelle générale et particulière des Crustacés et des Insectes,* vol. 3, 1–467. (For date of publication, see Dupuis, 1975).

Laubier, L. 1988. Le genre *Anomopsyllus* Sars, 1921, copepode parasite d'annelides polychetes: *A. pranizoides* Sars, 1921 et *A. abyssorum* nov. sp. *Crustaceana* 55:180–192.

Laubitz, D. R. 1993. Caprellidae (Crustacea: Amphipoda): towards a new synthesis. *Journal of Natural History* 27:965–976.

Leach, W. E. 1815. A tabular view of the external characters of four classes of animals, which Linn., arranged under Insecta; with the distribution of the genera composing three of these classes into orders, &c. and descriptions of several new genera and species. *Transactions of the Linnean Society of London* 11:306–400.

Lim, G. S. Y., and P. K. L. Ng. 1988. The first zoeal stage of *Harrovia albolineata* Adams and White, 1848 (Crustacea: Brachyura: Pilumnidae), with a note on Eumedonine systematics. *Journal of Natural History* 22:217–223.

Lincoln, R. J., and G. A Boxshall. 1983. Deep-sea asellote isopods of the north-east Atlantic: the family Dendrotionidae and some new ectoparasitic copepods. *Zoological Journal of the Linnean Society* 79:297–318.

Lopretto, E. C., and J J. Morrone. 1998. Anaspidacea, Bathynellacea (Crustacea, Syncarida), generalised tracks, and the biogeographical relationships of South America. *Zoologica Scripta* 27:311–318.

Lowry, J. K., and A. A. Myers. 2000. A family level phylogeny of iphemedioid amphipods (Crustacea, Amphipoda). *Abstracts of the 10th Colloqium on Amphipoda* (http://www.odu.edu/%7Ejrh100f/amphome): 14.

Lowry, J. K., and H. E. Stoddart. 1983. The shallow-water gammaridean Amphipoda of the subantarctic islands of New Zealand and Australia: Lysianassoidea. *Journal of the Royal Society of New Zealand* 13:279–394.

———. 1990. The Wandinidae, a new Indo-Pacific family of lysianassoid Amphipoda (Crustacea). *Records of the Australian Museum* 42:159–171.

———. 1995. The Amphipoda (Crustacea) of Madang Lagoon: Lysianassidae, Opisidae, Uristidae, Wandinidae and Stegocephalidae. In *Amphipoda (Crustacea) of the Madang Lagoon, Papua New Guinea,* ed. J. K. Lowry. *Records of the Australian Museum,* 22(supplement):97–174.

———. 1996. New lysianassoid amphipod species from Namibia and Madagascar (Lysianassidae Dana, 1849, Podoprionidae Fam. Nov.). *Bolletino di Museo Civico di Storia Naturale di Verona* 20:225–247.

———. 1997. Amphipoda Crustacea IV. Families Aristiidae, Cyphocarididae, Endevouridae, Lysianassidae, Scopelocheiridae, Uristidae. *Memoirs of the Hourglass Cruises* (Florida Marine Research Institute, St. Petersburg, Florida) 10:1–148.

Lützen, J., and T. Takahashi. 1996. Morphology and biology of *Polysaccus japonicus* (Crustacea, Rhizocephala, Akentrogonida, Polysaccidae, fam. N.), a parasite of the ghost-shrimp *Callianassa japonica. Zoologica Scripta* 25:171–181.

Maddison, W. P. 1997. Gene trees in species trees. *Systematic Biology* 46:523–536.

Magalhães, C., and M. Türkay. 1996a. Taxonomy of the Neotropical freshwater crab family Trichodactylidae. I. The generic system with description of some new genera (Crustacea: Decapoda: Brachyura). *Senckenbergiana Biologica* 75:63–95.

———. 1996b. Taxonomy of the Neotropical freshwater

crab family Trichodactylidae. II. The genera *Forsteria, Melocarcinus, Sylviocarcinus* and *Zilchiopsis* (Crustacea: Decapoda: Brachyura). *Senckenbergiana Biologica* 75:97–130.

———. 1996c. Taxonomy of the Neotropical freshwater crab family Trichodactylidae. III. The genera *Fredilocarcinus* and *Goyazana* (Crustacea: Decapoda: Brachyura). *Senckenbergiana Biologia* 75:131–142.

Maley, L. E., and C. R. Marshall. 1998. The coming of age of molecular systematics. *Science* 279:505–506.

Manning, R. B. 1995. Stomatopod Crustacea of Vietnam: the legacy of Raoul Serene. The Carcinological Society of Japan. *Crustacean Research,* Special no. 4: viii + 339 pp.

Manning, R. B., and A. J. Bruce. 1984. *Erythrosquilla megalops,* a remarkable new stomatopod from the western Indian Ocean. *Journal of Crustacean Biology* 4:329–332.

Manning, R. B., and D. K. Camp. 1993. Erythrosquilloidea, a new superfamily, and Tetrasquillidae, a new family of stomatopod crustaceans. *Proceedings of the Biological Society of Washington* 106:85–91.

Manning, R. B., and D. L. Felder. 1991. Revision of the American Callianassidae (Crustacea: Decapoda: Thalassinidea). *Proceedings of the Biological Society of Washington* 104:764–792.

———. 1996. *Nannotheres moorei,* a new genus and species of minute pinnotherid crab from Belize, Caribbean Sea (Crustacea: Decapoda: Pinnotheridae). *Proceedings of the Biological Society of Washington* 109:311–317.

Manning, R. B., and L. B. Holthuis. 1981. West African brachyuran crabs. *Smithsonian Contributions to Zoology* 306:1–379.

Marchenkov, A. V., and G. A. Boxshall. 1995. A new family of copepods associated with ascidiaceans in the White Sea, and an analysis of antennulary segmentation and setation patterns in the order Poecilostomatoida. *Zoologisches Anzeiger* 234:133–143.

Margolis, L., T. E. McDonald, and E. L. Bousfield. 2000. The whale-lice (Amphipoda: Cyamidae) of the northeastern Pacific region. *Amphipacifica* 2:63–117.

Marincek, M., and B. Petrov. 1991. A review of the Conchostraca (Crustacea) of Yugoslavia. *Bulletin du Muséum d'Histoire Naturelle de Belgrade,* series B, *Sciences Biologiques* 39:105–122.

Martens, K. 1992. On *Namibcypris costata* n. gen., n. sp. (Crustacea, Ostracoda, Candoninae) from a spring in northern Namibia, with the description of a new tribe and a discussion on the classification of the Podocopina. *Stygologia* 7:27–42.

Martens, K., D. J. Horne, and H. I. Griffiths. 1998. Age and diversity of non-marine ostracodes. In *Sex and parthenogenesis: evolutionary ecology of reproductive modes in non-marine ostracods,* ed. K. Martens, 37–55. Leiden: Backhuys Publishers, 336 pp.

Martin, J. W. 1984. Notes and bibliography on the larvae of xanthid crabs, with a key to the known zoeas of the western Atlantic and Gulf of Mexico. *Bulletin of Marine Science* 34:220–239.Martin, J. W. 1988. Phylogenetic significance of the brachyuran megalopa: evidence from the Xanthidae. *Symposium of the Zoological Society of London* 59:69–102.

———. 1991. Crabs of the family Homolodromiidae, III. First record of the larvae. *Journal of Crustacean Biology* 11:156–161.

———. 1992. Branchiopoda. In *Microscopic anatomy of invertebrates,* vol. 9, *Crustacea,* ed. F. W. Harrison and A. G. Humes, 25–224. New York: Wiley-Liss.

———. 1995. Review of: Walossek, D. 1993. The Upper Cambrian *Rehbachiella* and the phylogeny of Branchiopoda and Crustacea. *Earth-Science Reviews* 38: 72–75.

———. 1996. The Crustacean biodiversity survey. *Program and Abstracts, The Crustacean Society 1996 Summer Meeting, San Diego, California*: 28.

Martin, J. W., and L. G. Abele. 1986. Phylogenetic relationships of the genus *Aegla* (Decapoda: Anomura: Aeglidae), with comments on anomuran phylogeny. *Journal of Crustacean Biology* 6:576–616.

Martin, J. W., and D. Belk. 1988. Review of the clam shrimp family Lynceidae Stebbing, 1902 (Branchiopoda: Conchostraca), in the Americas. *Journal of Crustacean Biology* 8:451–482.

Martin, J. W., and C. Cash-Clark. 1995. The external morphology of the onychopod 'cladoceran' genus *Bythotrephes* (Crustacea, Branchiopoda, Onychopoda, Cercopagididae), with notes on the morphology and phylogeny of the order Onychopoda. *Zoologica Scripta* 24:61–90.

Martin, J. W., and J. C. Christiansen. 1995a. A morphological comparison of the phyllopodous thoracic limbs of a leptostracan (*Nebalia* sp.) and a spinicaudate conchostracan (*Leptestheria* sp.), with comments on the use of Phyllopoda as a taxonomic category. *Canadian Journal of Zoology* 73:2283–2291.

———. 1995b. A new species of the shrimp genus *Chorocaris* Martin and Hessler, 1990 (Crustacea: Decapoda: Bresiliidae) from hydrothermal vent fields along the Mid-Atlantic Ridge. *Proceedings of the Biological Society of Washington* 108:220–227.

Martin, J. W., and S. Trautwein. In press. Fossil crabs (Crustacea, Decapoda, Brachyura) from Lothagam. In *Lothagam: the dawn of humanity in eastern Africa,* ed. M. G. Leakey and J. M. Harris. New York: Columbia University Press.

Martin, J. W., F. M. Truesdale, and D. L. Felder. 1985. Larval development of *Panopeus bermudensis* Benedict and Rathbun, 1891 (Brachyura, Xanthidae) with notes on zoeal characters in xanthid crabs. *Journal of Crustacean Biology* 5:84–105.

Martin, J. W., E. W. Vetter, and C. E. Cash-Clark. 1996. Description, external morphology, and natural history observations of *Nebalia hessleri,* new species (Phyllocarida: Leptostraca), from southern California, with a key to the extant families and genera of the Leptostraca. *Journal of Crustacean Biology* 16: 347–372.

Mattern, D., and M. Schlegel. 2001. Molecular evolution of the small subunit ribosomal DNA in woodlice (Crustacea, Isopoda, Oniscidea) and implications for oniscidean phylogeny. *Molecular Phylogenetics and Evolution* 18:54–65.

Mayrat, A., and M. de Saint Laurent. 1996. Classe des Malacostracés (Malacostraca Latreille, 1802). Considerations sur la classe des Malacostracés. In *Traité de Zoologie. Anatomie, Systématique, Biologie. Crustacés. Tome VII, Fascicule II. Généralités (suite) et Systématique,* ed. J. Forest, 841–863. Paris: Masson, 1002 pp.

McHugh, D., and K. M. Halanych. 1998. Introduction to the symposium: evolutionary relationships of Metazoan phyla: advances, problems, and approaches. *American Zoologist* 38:813–817.

McKenzie, K. G. 1991. Crustacean evolutionary sequences and consequences. In *Proceedings of the 1990 International Crustacean Conference*, ed. P. J. F. Davie and R. H. Quinn. *Memoirs of the Queensland Museum* 31:19–38.

McKenzie, K. G., K. J. Müller, and M. N. Gramm. 1983. Phylogeny of Ostracoda. In *Crustacean issues 1. Crustacean phylogeny,* ed. F. R. Schram, 29–46. Rotterdam: A. A. Balkema Press, 372 pp.

McKinnon, A. D. 1994. *Australopsyllus fallaz* gen. et sp. nov., the third known species of the family Thaumatopsyllidae (Copepoda: Cyclopoida). *Sarsia* 79: 27–32.

McLaughlin, P. A. 1983a. A review of the phylogenetic position of the Lomidae (Crustacea: Decapoda: Anomala). *Journal of Crustacean Biology* 3:431–437.

———. 1983b. Hermit crabs—are they really polyphyletic? *Journal of Crustacean Biology* 3:608–621.

McLaughlin, P. A., and P. Hogarth. 1998. Hermit crabs (Decapoda: Anomura: Paguridea) from the Seychelles. *Zoologisches Mededelingen* 318:1–48.

McLaughlin, P. A., and L. B. Holthuis. 1985. Anomura versus Anomala. *Crustaceana* 49:204–209.

McLaughlin, P. A., and R. Lemaitre. 1997. Carcinization in the Anomura—fact or fiction? I. Evidence from adult morphology. *Contributions in Zoology (Amsterdam)* 67:79–123.

———. 2000. Aspects of evolution in the anomuran superfamily Paguroidea: one larval prospective. *Invertebrate Reproduction and Development* 38:159–169.

McLay, C. L. 1991. Crustacea Decapoda: the sponge crabs (Dromiidae) of New Caledonia and the Philippines with a review of the genera. In *Résultats des Campagnes Musorstom,* ed. A. Crosnier. *Mémoires de Muséum National d'Histoire Naturelle,* vol. 10, no. 156, 11–251.

———. 1999. Crustacea Decapoda: revision of the family Dynomenidae. In *Résultats des Campagnes Musorstom,* ed. A. Crosnier. *Mémoires de Muséum National d'Histoire Naturelle,* vol. 20, no. 180, 427–569.

Meier, R., and S. Richter. 1992. Suggestions for a more precise usage of proper names of taxa. *Zeitschrift für Zoologische Systematik und Evolutionsforschung* 30:81–88.

Milius, S. 1999. Should we junk Linnaeus? *Science News* 156:268–270.

Min, G. S., S. H. Kim, and W. Kim. 1998. Molecular phylogeny of arthropods and their relatives: polyphyletic origin of arthropodization. *Molecules and Cells* 8: 75–83.

Mizrahi, L., Y. Achituv, D. J. Katcoff, and R. Perl-Treves. 1998. Phylogenetic position of *Ibla* (Cirripedia: Thoracica) based on 18S rDNA sequence analysis. *Journal of Crustacean Biology* 18:363–368.

Monod, Th. 1984. La position systématique des Thermosbaenacea. *Annales de Societe Royale de Zoologie Belgique* 114(supplement 1):204–206.

Monod, Th., and Ph. Cals. 1988. Systématique et évolution des Thermosbénacés (Arthropodes, Crustacés), d'après l'ordonnance des structures épidermiques superficielles. *Comptes Rendus Hebdomadaires des Seances de l'Academie des Sciences,* series 3, 306:99–108.

———. 1999. Ordre des Thermosbaenacés (Thermosbaenacea Monod, 1927). In *Traité de Zoologie. Anatomie, Systématique, Biologie. Tome VII, Fascicule IIIA. Crustacés Péracarides,* ed. J. Forest. Memoires de l'Institut Oceanographique Fondation Albert Ier, Prince de Monaco, 19:11–34.

Monod, Th., and J. Forest. 1996. Histoire de la classification des Crustacés. In *Traité de Zoologie. Anatomie, Systématique, Biologie. Crustacés. Tome VII, Fascicule II. Généralités (suite) et Systématique,* ed. J. Forest, 235–267. Paris: Masson, 1002 pp.

Monod, Th., and L. Laubier. 1996. Les Crustacés dans la Biosphere. In *Traité de Zoologie. Anatomie, Systématique, Biologie. Crustacés. Tome VII, Fascicule II. Généralités (suite) et Systématique,* ed. J. Forest, 91–166. Paris: Masson, 1002 pp.

Moore, R. C. 1969. Editorial preface. In *Treatise on invertebrate paleontology, part R, Arthropoda 4,* ed. R. C. Moore, xi–xxxvi. Lawrence, Kansas: Geological Society of America and the University of Kansas Press.

Moore, R. C., and McCormick. 1969. General features of Crustacea. In *Treatise on invertebrate paleontology, part R, Arthropoda 4,* ed. R. C. Moore, R57—R120. Lawrence, Kansas: Geological Society of America and the University of Kansas Press.

Moosa, M. K. 1991. The Stomatopoda of New Caledonia and Chesterfield Islands. In *Le benthos des fonds meubles des lagons de Nouvelle-Calédonie,* vol. 1, ed. B. Richer de Forges, 149–219. Paris: ORSTOM Editions.

Morin, J. G., and A. C. Cohen. 1991. Bioluminescent displays, courtship, and reproduction in ostracodes. In *Crustacean sexual biology,* ed. R. T. Bauer and J. W. Martin, 1–16. New York: Columbia University Press.

Morrison, C. L., and C. W. Cunningham. 1999. Dramatic mitochondrial DNA gene rearrangements clarify relationships among anomuran crustaceans. *Program and Abstracts, The Crustacean Society 1999 Summer Meeting, Lafayette, Louisiana*: 40.

Moura, G., and M. L. Christoffersen. 1996. The system of the mandibulate arthropods: Tracheata and Remipedia as sister groups, "Crustacea" non-monophyletic. *Journal of Comparative Biology* 1:95–113.

Müller, H.-G. 1994. *World catalogue and bibliography of the Recent Stomatopoda.* Berlin: Wissenschaftlicher Verlag, 228 pp.

Müller, K. J. 1982. *Hesslandona unisulcata* sp. nov. (Ostracoda) with phosphatized appendages from Upper Cambrian 'Orsten' of Sweden. In *A research manual of fossil and Recent ostracods,* ed. R. H. Bate et al., 276–307. Chichester, England: Ellis Horwood.

———. 1983. Crustaceans with preserved soft parts from the Upper Cambrian of Sweden. *Lethaia* 16:93–109.

Müller, K. J., and D. Walossek. 1985a. A remarkable arthropod fauna from the Upper Cambrian "Orsten" of Sweden. *Transactions of the Royal Society of Edinburgh, Earth Sciences* 76:161–172.

———. 1985b. Skaracarida, a new order of Crustacea from the Upper Cambrian of Västergötland, Sweden. *Fossils and Strata* 17:1–65.

———. 1986a. *Martinssonia elongata* gen. et. sp. n., a crustacean-like euarthropod from the Upper Cambrian 'Orsten' of Sweden. *Zoologica Scripta* 15:73–92.

———. 1986b. Arthropod larvae from the Upper Cambrian of Sweden. *Transactions of the Royal Society of Edinburgh, Earth Sciences* 77:157–179.

———. 1988. External morphology and larval development of the Upper Cambrian maxillopod *Bredocaris admirabilis. Fossils and Strata* 23:1–70.

Negrea, S., N. Botnariuc, and H. J. Dumont. 1999. Phylogeny, evolution and classification of the Branchiopoda (Crustacea). *Hydrobiologia* 412:191–212.

Newman, W. A. 1982. Evolution within the Crustacea, part 3: Cirripedia. In *The biology of Crustacea, vol. 1,* Systematics, the fossil record, and biogeography, ed. L. G. Abele, 197–221. New York: Academic Press.

———. 1983. Origin of the Maxillopoda; urmalacostracan ontogeny and progenesis. In *Crustacean issues 1. Crustacean phylogeny,* ed. F. R. Schram, 105–119. Rotterdam: A. A. Balkema Press, 372 pp.

———. 1987. Evolution of cirripedes and their major groups. In *Crustacean issues 5. Barnacle biology,* ed. A. J. Southward, 3–42. Rotterdam: A. A. Balkema Press, 443 pp.

———. 1989. Juvenile ontogeny and metamorphosis in the most primitive living sessile barnacle, *Neoverruca,* from abyssal hydrothermal springs. *Bulletin of Marine Science* 45:467–477.

———. 1991. Cirripedia. In *Encyclopedia Britannica,* 15th edition, vol. 16, 849–854, 859. Chicago: Encyclopedia Britannica.

———. 1992. Origin of Maxillopoda. *Acta Zoologica* 73: 271–392.

———. 1993. Darwin and cirripedology. In *Crustacean issues 8. History of carcinology,* ed. F. M. Truesdale, 349–434. Rotterdam: A. A. Balkema Press, 445 pp.

———. 1996. Sous-classe de Cirripedes (Cirripedia Burmeister, 1834) super-ordre des Thoraciques et des Acrothoraciques (Thoracica Darwin, 1854—Acrothoracica Gruvel, 1905). In *Traité de Zoologie. Anatomie, Systématique, Biologie. Crustacés. Tome VII, Fascicule II. Généralités (suite) et Systématique,* ed. J. Forest, 453–540. Paris: Masson, 1002 pp.

Newman, W. A., and R. R. Hessler. 1989. A new abyssal hydrothermal verrucomorphan (Cirripedia; Sessilia): the most primitive living sessile barnacle. *Transactions of the San Diego Society of Natural History* 21:259–273.

Newman, W. A., and T. Yamaguchi. 1995. A new sessile barnacle (Cirripedia: Balanomorpha) from the Lau Back-Arc Basin, Tonga: first record of a living representative since the Miocene. *Bulletin du Museum National d'Histoire Naturelle (Paris)* 17A:211–243.

Ng, P. K. L. 1983. *Aspects of the systematics of the family Pilumnidae Samouelle, 1819 (Crustacea, Decapoda, Brachyura) and a study of evolutionary trends within the superfamily Xanthoidea (sensu Guinot, 1978).* Unpublished B. Sc. Honors Thesis, Department of Zoology, National University of Singapore, 251 pp, 12 plates.

———. 1988. *The freshwater crabs of peninsular Malaysia and Singapore.* Singapore: Department of Zoology, University of Singapore, and Shing Lee Publishers, 156 pp.

———. 1998. Marine crabs. In *FAO species identification guide for fishery purposes. The living marine resources of the Western-Central Pacific,* vol. 2. *Cephalopods, crustaceans, holothurians and sharks,* ed. K. E. Carpenter and V. H. Niem, 1045–1155. Rome: Food and Agriculture Organisation.

Ng, P. K. L, and P. F. Clark. 1999. The systematic position of the Eumedonidae: evidence from adult and larval morphology. In *General information and book of abstracts, 7th Colloquium Crustacea Decapoda Mediterranea,* ed. J. Paula. Lisbon, Portugal: University of Lisbon, 179 pp.

———. 2000. The eumedonid file: a case study of systematic compatibility using larval and adult characters (Crustacea: Decapoda: Brachyura). *Invertebrate Reproduction and Development* 38:225–252.

Ng, P. K. L., and G. Rodriguez. 1986. New records of *Mimilambrus wileyi* Williams, 1979 (Crustacea: Decapoda: Brachyura), with notes on the systematics of the Mimilambridae Williams, 1979 and Parthenopoidea MacLeay, 1838 *sensu* Guinot, 1978. *Proceedings of the Biological Society of Washington* 99: 88–99.

Ng, P. K. L., and B. Sket. 1996. The freshwater crab fauna (Crustacea: Brachyura) of the Philippines. IV. On a collection of Parathelphusidae from Bohol. *Proceedings of the Biological Society of Washington* 109: 695–706.

Ng, P. K. L., and M. Takeda. 1994. *Skelosothelphusa* (Crustacea, Decapoda, Brachyura), a new genus of potamonautid freshwater crab from Madagascar, with descriptions of two new species. *Bulletin of the National Science Museum, Tokyo, series A (Zoology)* 20:161–172.

Nicholls, G. E. 1943. The Phreatoicoidea: part I. The Amphisopidae. *Papers and Proceedings of the Royal Society of Tasmania,* 1942:1–145.

———. 1944. The Phreatoicoidea: part II. The Phreatoicidae. *Papers and Proceedings of the Royal Society of Tasmania,* 1943:1–156.

Nicol, S., and Y. Endo. 1999. Krill fisheries: development, management and ecosystem implications. *Aquatic Living Resources* 12:105–120.

Nielsen, C. 1995. *Animal evolution: interrelationships of the living phyla.* Oxford: Oxford University Press, 467 pp.

Nielsen, C., N. Scharff, and D. Eibye-Jacobsen. 1996. Cladistic analysis of the animal kingdom. *Biological Journal of the Linnean Society* 57:385–410.

Nielsen, C., W. F. Walker, H. R. Bode, and R. E. Steele. 1989. Phylogeny and molecular data (published as three separate comments on the paper by Field et al., 1988). *Science* 243:548–551.

Nixon, K., and J. Carpenter. 2000. On the other "phylogenetic systematics." *Cladistics* 16:298–318.

Nouvel, H., J.-P. Casanova, and J.-P. Lagardère. 1999. Ordre des Mysidacés (Mysidacea Boas 1883). In *Traité de Zoologie. Anatomie, Systématique, Biologie. Tome VII, Fascicule IIIA. Crustacés Péracarides,* ed. J. Forest. Memoires de l'Institut Oceanographique Fondation Albert Ier, Prince de Monaco, 19:39–86.

Ohtsuka, S., H. S. J. Roe, and G. A. Boxshall. 1993. A new family of calanoid copepods, the Hyperbionychidae, collected from the deep-sea hyperbenthic community in the northeastern Atlantic. *Sarsia* 78: 69–82.

Olesen, J. 1996. External morphology and phylogenetic significance of the dorsal/neck organ in the Conchostraca and the head pores of the cladoceran family Chydoridae (Crustacea, Branchiopoda). *Hydrobiologia* 330:213–226.

———. 1998. A phylogenetic analysis of the Conchostraca and Cladocera (Crustacea, Branchiopoda, Diplostraca). *Zoological Journal of the Linnean Society* 122:491–536.

———. 1999a. Larval and post-larval development of the branchiopod clam shrimp *Cyclestheria hislopi* (Baird, 1859) (Crustacea, Branchiopoda, Conchostraca, Spinicaudata). *Acta Zoologica* 80:163–184.

———. 1999b. A new species of *Nebalia* (Crustacea, Leptostraca) form Unguja Island (Zanzibar), Tanzania, East Africa, with a phylogenetic analysis of leptostracan genera. *Journal of Natural History* 33:1789–1809.

———. 2000. An updated phylogeny of the Conchostraca–Cladocera clade (Branchiopoda, Diplostraca). *Crustaceana* 73:869–886.

———. 2001. External morphology and larval development of *Derocheilocaris remanei* Delamare-Deboutteville & Chapuis, 1951 (Crustacea, Mystacocarida), with a comparison of crustacean segmentation and tagmosis patterns. *Biologiske Skrifter* 53:1–59.

Olesen, J., J. W. Martin, and E. Roessler. 1997. External morphology of the male of *Cyclestheria hislopi* (Baird, 1859) (Crustacea, Branchiopoda, Spinicaudata), with comparison of male claspers among the Conchostraca and Cladocera and its bearing on phylogeny of the "bivalved" Branchiopoda. *Zoologica Scripta* 25:291–316.

Orr, P. J., and D. E. G. Briggs. 1999. Exceptionally preserved conchostracans and other crustaceans from the Upper Carboniferous of Ireland. *Special Papers in Paleontology* (The Paleontology Association of London) 62:1–68.

Overstreet, R. M., I. Dyková, and W. E. Hawkins. 1992. Branchiura. In *Microscopic anatomy of invertebrates,* vol. 9, *Crustacea,* ed. F. W. Harrison and A. G. Humes, 385–413. New York: Wiley-Liss.

Page, R. D. M., and M. A. Charleston. 1998. From gene to organismal phylogeny: reconciled trees and the gene tree/species tree problem. *Molecular Phylogenetics and Evolution* 7:231–240.

Palmer, A. R. 1993. Whither Pentastomida? *Nature* 361: 214.

Panganiban, G., S. M. Irvine, C. Lowe, H. Roehl, L. S. Corley, B. Sherbon, J. K. Grenier, J. F. Fallon, J. Kimble, M. Walker, G. A. Wray, B. J. Swalla, M. Q. Martindale, and S. B. Carroll. 1997. The origin and evolution of animal appendages. *Proceedings of the National Academy of Sciences* 94:5162–5166.

Panganiban, G., A. Sebring, L. Nagy, and S. Carroll. 1995. The development of crustacean limbs and the evolution of arthropods. *Science* 270:1363–1366.

Park, T. 1986. Phylogeny of calanoid copepods. *Syllogeus* 58:191–196.

Paul, D. H. 1989. A neurophylogenists view of decapod Crustacea. *Bulletin of Marine Science* 45:487–504.

Pennant, T. 1777. *The British zoology,* vol. 4, 4th edition, *Crustacea, Mollusca, Testacea.* London: B. White, 136 pp.

Pérez Farfante, I., and B. F. Kensley. 1997. Penaeoid and sergestoid shrimps and prawns of the world. Keys and diagnoses for the families and genera. *Mémoires du Muséum National d'Histoire Naturelle,* vol. 175, 1–233.

Perl-Treves, R., L. Mizrahi, D. J. Katcoff, and Y. Achituv. 2000. Elucidation of the phylogenetic relationship of three thecostracans, *Verruca, Paralepas,* and *Dendrogaster* based on 18S rDNA sequence. *Journal of Crustacean Biology* 20:385–392.

Pinna, G., P. Arduini, C. Pesarini, and G. Teruzzi. 1982. Thylacocephala: una nuova classe di Crostacei fossili. *Atti della Societá Italiano di Scienze Naturali e del Museo Civico di Storia Naturale Milano* 123: 469–482.

———. 1985. Some controversial aspects of the morphology and anatomy of *Ostenocaris cypriformis* (Crustacea, Thylacocephala). *Transactions of the Royal Society of Edinburgh* 76:373–379.

Pires, A. M. S. 1987. *Potiicoara brasiliensis*: a new genus and species of Spelaeogriphacea (Crustacea: Peracarida) from Brazil with a phylogenetic analysis of the Peracarida. *Journal of Natural History* 21:225–238.

Pirlot, J. M. 1934. Les amphipodes de l'expedition du Siboga. Deuxieme partie, II. Les amphipodes de lar mer profonde. 2. Hyperiopsidae, Jassidae. *Siboga Expedite* 33d:167–235.

Pleijel, F., and G. W. Rouse. 2000. Least-inclusive taxonomic unit: a new taxonomic concept for biology. *Proceedings of the Royal Society of London* 267B: 627–630.

Pohle, G., and F. Marques. Larval stages of *Paradasygyius depressus* (Bell, 1835) (Crustacea: Decapoda: Brachyura: Majidae) and a phylogenetic analysis for 21 genera of Majidae. *Proceedings of the Biological Society of Washington* 113:739–760.

Poore, G. C. B. 1987. Lynseiidae (Isopoda, Flabellifera), a new monotypic family from Australia. *Journal of Crustacean Biology* 7:258–264.

———. 1994. A phylogeny of the families of Thalassinidea (Crustacea: Decapoda) with keys to families and genera. *Memoirs of the Museum of Victoria* 54:79–120.

———. 1997. A review of the thalassinidean families Callianideidae Kossman, Micheleidae Sakai, and Thomassiniidae de Saint Laurent (Crustacea, Decapoda) with descriptions of fifteen new species. *Zoosystema* 19:345–420.

———. 1998. Phylogeny of the Anthuridea (Isopoda). *Proceedings and Abstracts of the 4th International Crustacean Congress, Amsterdam*: 71 (abstract 42).

———. 2001a. Families and genera of Isopoda Anthuridea. In *Crustacean issues 13. Isopod systematics and evolution,* ed. B. Kensley and R. C. Brusca, 63–174. Rotterdam: A. A. Balkema Press, 357 pp.

———. 2001b. Isopoda Valvifera: diagnoses and relationships of the families. *Journal of Crustacean Biology* 21:205–230.

Poore, G. C. B., and T. M. Bardsley. 1992. Austrarcturellidae (Crustacea Isopoda: Valvifera), a new family from Australia. *Invertebrate Taxonomy* 6:843–908.

Poore, G. C. B., and W. F. Humphreys. 1998. First record of Spelaeogriphacea from Australasia: a new genus and species from an aquifer in the arid Pilbara of Western Australia. *Crustaceana* 71:721–742.

Poore, G. C. B., and H. M. Lew Ton. 1988. Antheluridae, a new family of Crustacea (Isopoda: Anthuridea) with new species from Australia. *Journal of Natural History* 22:489–506.

Popadić, A., D. Rusch, M. Peterson, B. T. Rogers, and T. C. Kaufman. 1996. Origin of the arthropod mandible. *Nature* 380:395.

Por, F. D. 1984. Canuellidae Lang (Harpacticoida: Polyarthra) and the ancestry of the Copepoda. *Crustaceana* 7(supplement):1–24.

———. 1986. A re-evaluation of the family Cleotodidae Sars, Lang (Copepoda, Harpacticoida). *Syllogeus* 58: 420–425.

Pretzmann, G. 1973. Grundlagen und Ergebnisse der Systematik der Pseudothelphusidae. *Zeitschrift für Zoologische Systematik und Evolutionsforschung* 11:196–218.

Preuss, G. 1951. Die Verwandtschaft der Anostraca und Phyllopoda. *Zoologisches Anzeiger* 147:50–63.

Raff, R. A., C. R. Marshall, and J. M. Turbeville. 1994.

Using DNA sequences to unravel the Cambrian radiation of the animal phyla. *Annual Review of Ecology and Systematics* 25:351–375.

Rafinesque, C. S. 1815. *Analyse de la nature ou tableau de l'univers et des corps organisés.* Palermo: l'Imprimerie de Jean Barravecchia, 224 pp.

Raibaut, A. 1996. Copépodes II. Les Copépodes parasites. In *Traité de Zoologie. Anatomie, Systématique, Biologie. Crustacés. Tome VII, Fascicule II. Généralités (suite) et Systématique,* ed. J. Forest, 639–718. Paris: Masson, 1002 pp.

Razouls, C. 1996. Copépodes I. Les Copépodes libres. In *Traité de Zoologie. Anatomie, Systématique, Biologie. Crustacés. Tome VII, Fascicule II. Généralités (suite) et Systématique,* ed. J. Forest, 571–638. Paris: Masson, 1002 pp.

Razouls, C., and A. Raibaut. 1996. Copépodes III. Phylogénie et classification. In *Traité de Zoologie. Anatomie, Systématique, Biologie. Crustacés. Tome VII, Fascicule II. Généralités (suite) et Systématique,* ed. J. Forest, 719–738. Paris: Masson, 1002 pp.

Regier, J. C., and J. W. Shultz. 1997. Molecular phylogeny of the major arthropod groups indicates polyphyly of crustaceans and a new hypothesis for the origin of hexapods. *Molecular Biology and Evolution* 14: 902–913.

———. 1998a. Resolving arthropod phylogeny using multiple nuclear genes. *American Zoologist* 37: 102A.

———. 1998b. Molecular phylogeny of arthropods and the significance of the Cambrian "explosion" for molecular systematics. *American Zoologist* 38:918–928.

Remigio, E. A., and P. D. Hebert. 2000. Affinities among anostracan (Crustacea: Branchiopoda) families inferred from phylogenetic analyses of multiple gene sequences. *Molecular Phylogenetics and Evolution* 17:117–128.

Ren, X.-Q. 1999. A new family of superfamily Haustorioidea (Crustacea: Amphipoda: Gammaridea) from the China sea. *Chinese Journal of Oceanology and Limnology* 17:344–349.

Rice, A. L. 1980. Crab zoeal morphology and its bearing on the classification of the Brachyura. *Transactions of the Zoological Society of London* 35:271–424.

———. 1981. The megalopa stage in brachyuran crabs. The Podotremata Guinot. *Journal of Natural History* 15:1003–1011.

———. 1983. Zoeal evidence for brachyuran phylogeny. In *Crustacean issues 1. Crustacean phylogeny,* ed. F. R. Schram, 313–329. Rotterdam: A. A. Balkema Press, 372 pp.

———. 1988. The megalopa stage in majid crabs, with a review of spider crab relationships based on larval characters. *Symposium of the Zoological Society of London* 59:27–46.

Richer de Forges, B., B. G. M. Jamieson, D. Guinot, and C. C. Tudge. 1997. Ultrastructure of the spermatozoan of Hymenosomatidae (Crustacea: Brachyura) and the relationships of the family. *Marine Biology* 130:233–242.

Richter, S. 1994. Monophylie der Lophogastrida und Phylogenetisches System der Peracarida (Crustacea). *Verhandlungen der Deutschen Zoologischer Gesellschaft* 87:231.

———. 1999. The structure of the ommatidia of the Malacostraca (Crustacea)—a phylogenetic approach. *Verhandlungen der Naturwissenschaftlichen Vereine in Hamburg* 38:161–204.

Richter, S., A. Braband, N. Aladin, and G. Scholtz. 2001. The phylogenetic relationships of "predatory water-fleas" (Cladocera: Onychopoda, Haplopoda) inferred from 12S rDNA. *Molecular Phylogenetics and Evolution* 18:1–9.

Richter, S., and G. Scholtz. 1994. Morphological evidence for a hermit crab ancestry of lithodids (Crustacea, Decapoda, Anomala, Paguroidea). *Zoologische Anzeiger* 233:187–210.

———. In press. Phylogenetic analysis of the Malacostraca (Crustacea). *Journal of Zoological Systematics and Evolutionary Research.*

Riley, J. 1986. The biology of pentastomids. In *Advances in parasitology,* vol. 25, 45–128. London and New York: Academic Press.

Riley, J., A. A. Banaja, and J. L. James. 1978. The phylogenetic relationships of the Pentastomida: the case for their inclusion within the Crustacea. *International Journal for Parasitology* 8:245–254.

Rodríguez, G. 1982. *Les crabes d'eau douce d'Amerique. Familie des Pseudothelphusidae.* Paris: ORSTOM, Faune Tropicale, vol. 22, 223 pp.

———. 1986. Centers of radiation of freshwater crabs in the Neotropics. In *Crustacean issues 4. Crustacean biogeography,* ed. R. H. Gore and K. L. Heck, 51–67. Rotterdam: A. A. Balkema Press, 292 pp.

———. 1992. *The freshwater crabs of America. Family Trichodactylidae and a supplement to the family Pseudothelphusidae.* Paris: ORSTOM, Faune Tropicale, vol. 31, 189 pp.

Roessler, E. W. 1991. Estudios sobre Entomostraces de Colombia VI—Paraimnadiidae, una nueva familia de Crustacea—Conchostraca. *Revista de la Academia Colombiana de Ciencias Exactas, Físicas y Naturales* 18:93–104.

———. 1995a. Review of Colombian conchostraca (Crustacea)—morphotaxonomic aspects. *Hydrobiologia* 298 (*Developmental Hydrobiology* 103):253–262.

———. 1995b. Review of Colombian conchostraca (Crustacea)—ecological aspects and life cycles—families Lynceidae, Limndadiidae, Leptestheriidae, and Metalimnadiidae. *Hydrobiologia* 298 (*Developmental Hydrobiology* 103):113–124.

Rolfe, W. D. I. 1981. Phyllocarida and the origin of the Malacostraca. *Geobios* 14:17–27.

———. 1985. Form and function in the Thylacocephala, Conchyliocarida and Concavicarida (?Crustacea): a problem of interpretation. *Transactions of the Royal Society of Edinburgh, Earth Sciences,* 76:391–399.

———. 1992. Not yet proven Crustacea: the Thylacocephala. *Acta Zoologica* 73:301–304.

Roman, M.-L., and H. Dalens. 1999. Ordre des Isopodes (Épicarides exclus) (Isopoda Latreille, 1817). In *Traité de Zoologie. Anatomie, Systématique, Biologie. Tome VII, Fascicule IIIA. Crustacés Péracarides,* ed. J. Forest. Memoires de l'Institut Oceanographique Fondation Albert Ier, Prince de Monaco, 19:177–278.

Roush, W. 1995. Gene ties arthropods together. *Science* 270:1297.

Rudolphi, K. A. 1809. Entozoorum, sive Vermium Intestinalis Historia naturalis 2 (1).

Ruppert, E. E., and R. D. Barnes. 1994. *Invertebrate zoology,* 6th edition. Orlando, Florida: Saunders College Publishing and Harcourt Brace Jovanovich, 1056 pp.

Saint Laurent, M. de. 1979. Vers une nouvelle classification des Crustaces Decapodes Reptantia. *Bulletin de l'Office National des Pêches République Tunisienne (Tunisia), Ministere de l'Agriculture* 3:15–31.

———. 1980a. Sur classification et phylogenie des Crustaces Decapodes brachyoures. I. Podotremata Guinot, 1977, et Eubrachyura sect. nov. *Comptes Rendus Hebdomadaires des Seances de l'Academie des Sciences,* series D, 290:1265–1268.

———. 1980b. Sur classification et phylogenie des Crustaces Decapodes brachyoures. II. Heterotremata et Thoracotremata Guinot, 1977. *Comptes Rendus Hebdomadaires des Seances de l'Academie des Sciences,* series D, 290:1317–1320.

———. 1985. Remarques sur la distribution des Crustacés Décapodes. In *Peuplements profonds du Golfe de Gascogne, Campagnes Biogas,* ed. L. Laubier and C. Monniot, 469–478. Ifremer: Institut Francaise de Rechereche pour l'Exploraion de la Mer.

———. 1988. Enoplometopoidea, nouvelle superfamille de Crustacés Décapodes Astacidea. *Comptes Rendus Hebdomadaires des Seances de l'Academie des Sciences,* series 3, 307:59–62.

———. 1989. La nouvelle superfamille des Retroplumoidea Gill, 1894 (Decapoda, Brachyura): systématique, affinités et évolution. In *Résultats des Campagnes MUSORSTOM,* vol. 5, ed. J. Forest. *Mémoires du Muséum National d'Histoire Naturelle (Paris),* sére A, vol. 144, 103–179.

Sakai, K. 1992. The families Callianideidae and Thalassinidae, with the description of two new subfamilies, one new genus and two new species. *Naturalists* 4: 1–33.

———. 1999. Synopsis of the family Callianassidae, with keys to subfamilies, genera and species, and the description of new taxa (Crustacea: Decapoda: Thalassinidea). *Zoologische Verhandelingen (Leiden)* 326:1–152.

Samouelle, G. 1819. *The entomologist's useful compendium; or an introduction to the knowledge of British insects, comprising the best means of obtaining and preserving them, and a description of the apparatus generally used; together with the genera of Linné. and the modern method of arranging the classes Crustacea, Myriapoda, Spiders, Mites and Insects, from their affinities and structure, according to the views of Dr. Leach, Also an explanation of the terms used in entomology; a calendar of the times of appearance and usual situations of near 3,000 species of British insects; with instructions for collecting and fitting up objects for the microscope.* London: Thomas Boys, 496 pp.

Sanders, H. L. 1963. Significance of the Cephalocarida. In *Phylogeny and evolution of Crustacea,* ed. H. B. Whittington and W. D. I Rolfe, 163–175. Cambridge, Massachusetts: Special Publication of the Museum of Comparative Zoology.

Sanders, H. L., R. R. Hessler, and S. P. Garner. 1985. *Hirsutia bathyalis,* a new unusual deep-sea benthic peracaridan crustacean from the tropical Atlantic. *Journal of Crustacean Biology* 5:30–57.

Sassaman, C. 1992. Description of the mature female and epicaridium lava of *Cabiropsis montereyensis* Sassaman from southern California (Crustacea: Isopoda: Cabiropidae). *Proceedings of the Biological Society of Washington* 105:575–584.

———. 1995. Sex determination and evolution of unisexuality in the Conchostraca. *Hydrobiologia* 298:45–65.

Schmalfuss, H. 1989. Phylogenetics in Oniscidea. *Monitore Zoologico Italiano, n. s., Monografia* 4:3–27.

Schmidt, M., and S. Harzsch. 1999. Comparative analysis of neurogenesis in the central olfactory pathway of adult decapod crustaceans by *in vivo* BrdU labeling. *Biological Bulletin* 196:127–136.

Schmidt-Rhaesa, A., U. Ehlers, T. Bartolomaeus, C. Lemburg, and J. R. Garey. 1998. The phylogenetic position of the Arthropoda. *Journal of Morphology* 238:263–285.

Schminke, H. K. 1973. Evolution, System und Verbreitungsgeschichte der Familie Parabathynellidae (Bathynellacea, Malacostraca). *Mikrofauna des Meeresbodens* 24:1–192.

Schmitt, W. L. 1965. *Crustaceans.* Ann Arbor, Michigan: The University of Michigan Press, 204 pp.

Scholtz, G. 1995. Expression of the *engrailed* gene reveals nine putative segment-anlagen in the embryonic pleon of the freshwater crayfish *Cherax destructor* (Crustacea, Malacostraca, Decapoda). *Biological Bulletin* 188:157–165.

———. 1998. Von Zellen und Kontinenten—die Evolution der Flußkrebse (Decapoda, Astacida). *Stapfia* 58:205–212.

———. 1999. Freshwater crayfish evolution. *Freshwater crayfish 12 (Proceedings of the Twelfth Symposium of the International Association of Astacology)*: 36–48.

Scholtz, G., and S. Richter. 1995. Phylogenetic systematics of the reptantian Decapoda (Crustacea, Malacostraca). *Zoological Journal of the Linnean Society* 113: 289–328.

Schram, F. R. 1971. Polyphyly in the Eumalacostraca? *Crustaceana* 16:243–250.

———. 1981. On the classification of Eumalacostraca. *Journal of Crustacean Biology* 1:1–10.

———. (editor). 1983a. *Crustacean issues 1. Crustacean phylogeny.* Rotterdam: A. A. Balkema Press, 372 pp.

———. 1983b. Remipedia and crustacean phylogeny. In *Crustacean issues 1. Crustacean phylogeny,* ed. F. R. Schram, 23–28. Rotterdam: A. A. Balkema Press, 372 pp.

———. 1984a. Relationships within eumalacostracan Crustacea. *Transactions of the San Diego Society of Natural History* 20:301–312.

———. 1984b. Fossil Syncarida. *Transactions of the San Diego Society of Natural History* 20:189–246.

———. 1986. *Crustacea.* New York: Oxford University Press, 606 pp.

———. 1991. Arthropod pattern theory: a new approach to arthropod phylogeny. In *Proceedings of the 1990 International Crustacean Conference,* ed. P. J. F. Davie and R. H. Quinn. *Memoirs of the Queensland Museum* 31:1–18.

Schram, F. R., and J. T. Høeg. 1995. New frontiers in barnacle evolution. In *Crustacean issues 10. New frontiers in barnacle evolution,* ed. F. R. Schram and J. T. Høeg, 297–312. Rotterdam: A. A. Balkema Press.

Schram, F. R., and C. H. J. Hof. 1998. Fossils and the interrelationships of major crustacean groups. In *Arthropod fossils and phylogeny,* ed. G. D. Edgecombe, 233–302. New York: Columbia University Press.

Schram, F. R., C. H. J. Hof, and F. A. Steeman. 1999. Thylacocephala (Arthropoda: Crustacea?) from the

Cretaceous of Lebanon and implications for thylacocephalan systematics. *Palaeontology* 42:769–797.

Schram, F. R., and C. Lewis. 1989. Functional morphology of feeding in the Nectiopoda. In *Crustacean issues 6. Functional morphology of feeding and grooming in Crustacea,* ed. B. E. Felgenhauer, L. Watling, and A. B. Thistle, 115–122. Rotterdam: A. A. Balkema Press.

Schram, F. R., J. Sieg, and E. Malzahn. 1986. Fossil Tanaidacea. *Transactions of the San Diego Society of Natural History* 21:127–144.

Schram, F. R., S. Yanbin, R. Vonk, and R. S. Taylor. 2000. The first fossil stenopodidean. *Crustaceana* 73:235–242.

Schram, F. R., R. Vonk, and C. H. J. Hof. 1997. Mazon Creek Cycloidea. *Journal of Paleontology* 71:261–284.

Schram, F. R., and J. C. von Vaupel Klein (editors). 1999. *Crustaceans and the biodiversity crisis. Proceedings of the 4th International Crustacean Congress, Amsterdam.* Leiden: Brill, 1021 pp.

Schram, F. R., J. Yager, and M. J. Emerson. 1986. Remipedia, part 1, systematics. *San Diego Society of Natural History Memoir* 15:1–60.

Schubart, C. D., J. A. Cuesta, R. Diesel, and D. L. Felder. 2000a. Molecular phylogeny, taxonomy, and evolution of nonmarine lineages within the American grapsoid crabs (Crustacea: Brachyura). *Molecular Phylogenetics and Evolution* 15:179–190.

Schubart, C. D., J. A. Cuesta, and D. L. Felder. In press. Glyptograpsidae, a new brachyuran family from Central America: larval and adult morphology and a molecular phylogeny of the Grapsoidea. *Journal of Crustacean Biology.*

Schubart, C. D., J. A. Cuesta, and A. Rodríguez. In press. Molecular phylogeny of the crab genus *Brachynotus* (Brachyura: Varunidae) based on the 16S rRNA gene. *Hydrobiologia.*

Schubart, C. D., J. E. Neigel, and D. L. Felder. 1998. The use of the mitochondrial 16S gene for the study of crustacean phylogenies and biogeography. *Proceedings and abstracts of the 4th International Crustacean Congress, Amsterdam*: 93 (abstract 172).

———. 2000b. Molecular phylogeny of mud crabs (Brachyura: Panopeidae) from the northwestern Atlantic and the role of morphological stasis and convergence. *Marine Biology* 137:1167–1174.

———. 2000c. Use of the mitochondrial 16S rRNA gene for phylogenetic and population studies of Crustacea. In *Crustacean issues 12. The biodiversity crisis and Crustacea,* ed. J. C. von Vaupel Klein and F. R. Schram, 817–830. Rotterdam: A. A. Balkema Press.

Schultz, G. A. 1995. Terrestrial isopod crustaceans (Oniscidea) from Paraguay with definition of a new family. *Revue Suisse de Zoologie* 102:387–424.

Schweitzer, C. E., and R. M. Feldmann. 2000. New species of calappid crabs from western North America and reconsideration of the Calappidae sensu lato. *Journal of Paleontology* 74:230–246.

Schweitzer, C. E., and E. W. Salva. 2000. First recognition of the Cheiragonidae (Decapoda) in the fossil record and comparison of the family with the Atelecyclidae. *Journal of Crustacean Biology* 20:285–298.

Schwenk, K., A. Sand, M. Boersma, M. Brehem, E. Mader, D. Offerhaus, and P. Spaak. 1998. Genetic markers, genealogies and biogeographic patterns in the Cladocera. *Aquatic Ecology* 32:37–51.

Scott, A. 1909. The Copepoda of the Siboga Expedition. Part I. Free-swimming, littoral and semi-parasitic Copepoda. *Siboga Expeditie Monograph* 29a:1–323, plates 1–69.

Secretan, S. 1985. Conchyliocarida, a class of fossil crustaceans: relationships to Malacostraca and postulated behaviour. *Transactions of the Royal Society of Edinburgh, Earth Sciences* 76:381–389.

———. 1998. The sella turcica of crabs and the endophragmal system of decapods. *Journal of Natural History* 32:2753–1767.

Segonzac, M., M. de Saint Laurent, and B. Casanova. 1993. L'enigma du comportement trophique des crevettes Alvinocarididae des sites hydrothermaux de la dorsale medio-atlantique. *Cahiers de Biologie Marine* 34:535–571.

Self, J. T. 1969. Biological relationships of the Pentastomida; a bibliography on the Pentastomida. *Experimental Parasitology* 24:63–119.

Serejo, C. S. 2000. A preliminary cladistic analysis of the superfamily Talitroidea (Amphipoda, Gammaridea). *Abstracts of the 10th Colloqium on Amphipoda* (http://www.odu.edu/%7Ejrh 100f/amphome): 6–7.

Serène, R. 1984. Crustacés Décapodes Brachyoures de l'Ocean Indien Occidental et de la Mer Rouge, Xanthoidea: Xanthidae et Trapeziidae. Avec un addendum par Crosnier, A.: Carpiliidae et Menippidae. *Faune Tropicale* 24:1–400.

Seridji, R. 1993. Descriptions of some planktonic larvae of the Calappidae (Crustacea: Decapoda: Brachyura). *Journal of Plankton Research* 15:437–453.

Serov, P. A., and G. D. F. Wilson. 1999. A revision of the Pseudojaniridae Wilson, with a description of a new genus of Stenetriidae Hansen (Crustacea: Isopoda: Asellota). *Invertebrate Taxonomy* 13:67–116.

Shank, T. M., M. B. Black, K. M. Halanych, R. A. Lutz, and R. C. Vrijenhoek. 1999. Miocene radiation of deep-sea hydrothermal vent shrimp (Caridea: Bresiliidae): evidence from mitochondrial cytochrome oxidase subunit I. *Molecular Phylogenetics and Evolution* 13:244–254.Shen, Y. 1984. Occurrence of Permian leaid conchostracans in China and its palaeogeographical significance. *Acta Paleontologica Sinica* 29:505–513.

Shen, Y. 1990. A new conchostracan genus from Lower Permian Tungziyan Formation, Fujian. *Acta Paleontologica Sinica* 29:309–314.

Shen, Y., R. S. Taylor, and F. R. Schram. 1998. A new spelaeogriphacean (Crustacea: Peracarida) from the Upper Jurassic of China. *Contributions to Zoology (Amsterdam)* 68:19–35.

Shubin, N., C. Tabin, and S. Carroll. 1997. Fossils, genes and the evolution of animal limbs. *Nature* 388:639–648.

Shultz, J. W., and J. C. Regier. 2000. Phylogenetic analysis of arthropods using two nuclear protein-encoding genes supports a crustacean + hexapod clade. *Proceedings of the Royal Society of London* 267B: 1011–1019.

Sieg, J. 1973. *Ein Beitrag zum Natürlichen System der Dikonophora Lang. (Textteil, Tafelteil).* Ph.D. Dissertation, University of Kiel, 298 pp.

———. 1976. Zum Natürlichen System der Tanaidacea Lang (Crustacea, Tanaidacea). *Zeitschrift für Zoologischer Systematik und Evolutionsforschung* 14: 177–198.

———. 1982. Uber ein "connecting link" in der Phylogenie der Tanaidomorpha (Tanaidacea). *Crustaceana* 43:65–77.

———. 1983a. Evolution of Tanaidacea. In *Crustacean issues 1. Crustacean phylogeny,* ed. F. R. Schram, 229–254. Rotterdam: A. A. Balkema Press, 372 pp.

———. 1983b. Tanaidacea. In *Crustaceorum catalogus pars 6,* ed. H. E. Gruner and L. B. Holthuis, 1–552. The Hague: W. Junk.

———. 1984. Neuere Erkenntnisse zum Natürliche System der Tanaidacea. Eine phylogenetische Studie. *Zoologica (Stuttgart)* 136:1–132.

———. 1986a. Crustacea Tanaidacea of the Antarctic and subanctarctic. 1. On material collected at Tierra del Fuego, Isla de los Estados, and the west coast of the Antarctic peninsula. *Biology of the Antarctic Seas* 18:1–180.

———. 1986b. Tanaidacea (Crustacea) von der Antarktis und Subantarktis. II. Tanaidacea gesammelt von Dr. J. W. Wägele während der Deutschen Antarktis Expedition 1983. *Mitteilungen aus dem Zoologischen Museum der Universität Keil* 2:1–80.

Siveter, D. J., and M. Williams. 1997. Cambrian bradoriid and phosphatocopid arthropods of North America. *Special Papers in Paleontology* (The Paleontological Association of London) 57:1–69.

Smirnov, N. N. 1992. The Macrothricidae of the world. *Guides to the Identification of Microinvertebrates of the Continental Waters of the World* 1:1–143.

Soh, H. Y., S. Ohtsuka, H. Imabayashi, and H.-L. Suh. 1999. A new deep-water calanoid copepod and the phylogeny of the genus *Nullosetigera* nom. nov. in the Nullosetigeridae nom. nov. (pro *Phyllopus*: Phyllopodidae) from Japanese waters. *Journal of Natural History* 33:1581–1602.

Sohn, I. G. 1988. Darwinulocopina (Crustacea, Podocopa), a new suborder proposed for nonmarine Palaeozoic to Holocene Ostracoda. *Proceedings of the Biological Society of Washington* 101:817–824.

Souza-Kury, L. A. 1998. Malacostraca—Peracarida. Isopoda. Oniscidea. In *Catalogue of Crustacea of Brazil,* ed. P. S. Young, 653–674. Rio de Janeiro: Museu Nacional.

Spears, T., and L. G. Abele. 1988. Molecular phylogeny of brachyuran crustaceans based on 18S rRNA nucleotide sequences. *American Zoologist* 28:2A.

———. 1997. Crustacean phylogeny inferred from 18S rDNA. In *Arthropod relationships,* ed. R. A. Fortey and R. H. Thomas, 169–187. Systematics Association Special Volume series 55. London: Chapman and Hall.

———. 1998. The role of 18S rDNA hypervariable regions in crustacean phylogeny. *Proceedings and Abstracts of the 4th International Crustacean Congress, Amsterdam*: 92 (abstract 171).

———. 1999a. Crustacean molecular systematics and phylogeny: the 18S rDNA chronicles. *Program and Abstracts, The Crustacean Society 1999 Summer Meeting, Lafayette, Louisiana*: 44.

———. 1999b. The phylogenetic relationships of crustaceans with foliaceous limbs: an 18S rDNA study of Branchiopoda, Cephalocarida, and Phyllocarida. *Journal of Crustacean Biology* 19:825–843.

———. 2000. Branchiopod monophyly and interordinal phylogeny inferred from 18S ribosomal DNA. *Journal of Crustacean Biology* 20:1–24.

Spears, T., L. G. Abele, and M. A. Applegate. 1994. Phylogenetic study of cirripedes and selected relatives (Thecostraca) based on 18S rDNA sequence analysis. *Journal of Crustacean Biology* 14:641–656.

Spears, T., L. G. Abele, and W. Kim. 1992. The monophyly of brachyuran crabs: a phylogenetic study based on 18S rRNA. *Systematic Biology* 41:446–461.

Spears, T., N. Cumberlidge, and L. G. Abele. 2000. A molecular-based phylogeny of the freshwater crabs. *Program and Abstracts, The Crustacean Society 2000 Summer Meeting, Puerto Vallarta, Mexico*: 46.

Spivak, E. D., and J. A. Cuesta. 2000. Larval development of *Cyrtograpsus affinis* (Dana) (Decapoda, Brachyura, Varunidae) from Rio de la Plata estuary, reared in the laboratory. *Scientia Marina* 64:29–47.

Starobogatov, Y. I. 1986. Systematics of Crustacea. *Zoologicheskiy Zhurnal* 65:1769–1781. [In Russian, with English summary]

———. 1988. Systematics of Crustacea. *Journal of Crustacean Biology* 8:300–311. (English translation by M. J. Grygier of Starobogatov, 1986, above).

Stebbing, T. R. R. 1899. Revision of Amphipoda (continued). *Annals and Magazine of Natural History,* series 7, 4:205–211

Sternberg, R. von. 1996. Carcinization as an underlying synapomorphy for the decapod crustacean taxon Meiura. *Evolutionary Theory* 11:153–162.

———. 1997. Cladistics of the freshwater crab family Trichodactylidae (Crustacea: Decapoda): appraising the reappraisal. *Journal of Comparative Biology* 2:49–62.

Sternberg, R. von, and N. Cumberlidge. 1999. A cladistic analysis of *Platythelphusa* A. Milne Edwards, 1887, from lake Tanganyika, East Africa (Decapoda: Potamoidea: Platythelphusidae) with comments on the phylogenetic position of the group. *Journal of Natural History* 33:493–511.

———. 2000a. Cladistic relationships of the true freshwater crabs (Decapoda: Eubrachyura: Gecarcinucicoidea, Potamoidea, Pseudothelphusoidea, and Trichodactylidae). *Program and Abstracts, The Crustacean Society 2000 Summer Meeting, Puerto Vallarta, Mexico*: 46.

———. 2000b. Taxic relationships within the Grapsidae MacLeay, 1838 (Crustacea: Decapoda: Brachyura). *Journal of Comparative Biology* 3:115–136.

———. 2001. On the heterotreme-thoracotreme distinction in the Eubrachyura de Saint-Laurent, 1980 (Decapoda, Brachyura). *Crustaceana* 74:321–338.

———. In press. Notes on the position of the true freshwater crabs within the brachyrhynchan Eubrachyura (Crustacea: Decapoda: Brachyura). *Hydrobiologia.*

Sternberg, R. von, N. Cumberlidge, and G. Rodríguez. 1999. On the marine sister groups of the freshwater crabs (Crustacea: Decapoda: Brachyura). *Journal of Zoological Systematics and Evolutionary Research* 37:19–38.

Števčić, Z. 1973. The systematic position of the family Raninidae. *Systematic Zoology* 22:625–632.

———. 1983. Revision of the Calappidae. *Australian Museum Memoir* 18:165–171.

———. 1988. The status of the family Cheiragonidae Ortmann, 1893. *International Journal of Marine Biology and Oceanography (Taranto, Italy)* 14:1–14.

———. 1994. Contribution to the re-classification of the family Majidae. *Periodicum Biologorum* 96:419–420.

———. 1995. Brachyuran systematics and the position of the family Raninidae reconsidered. *Arthropoda Selecta* 4:27–36.

———. 1998. Evolutionary arrangement of the brachyuran families together with a checklist. *Periodicum Biologorum* 100:101–104.

Števčić, Z., P. Castro, and R. H. Gore. 1988. Re-establishment of the family Eumedonidae Dana, 1853 (Crustacea: Decapoda: Brachyura). *Journal of Natural History* 22:1301–1324.

Števčić, Z., and R. H. Gore. 1982. Are the Oxyrhyncha a natural group? *Thalassia Jugoslavica* 17:1–16.

Stock, J. H. 1968. *Vectoriella marinovi,* un copépode nouveau, parasite d'une annélide polychète pontique. *Crustaceana* 1(supplement 10):186–192.

Storch, V. 1984. Pentastomida. In *Biology of the integument,* ed. J. Bereiter-Hahn, A. G. Matoltsky, and K. S. Richards, 709–713. Berlin: Springer Verlag.

Storch, V., and B. G. M. Jamieson. 1992. Further spermatological evidence for including the Pentastomida (tongue worms) in the Crustacea. *International Journal of Parasitology* 22:95–108.

Strausfeld, N. J. 1998. Crustacean–insect relationships: the use of brain characters to derive phylogeny among segmented invertebrates. *Brain, Behavior, and Evolution* 52:186–206.

Suarez-Morales, E., and T. M. Iliffe. 1996. New superfamily of Calanoida (Copepoda) from an anchialine cave in the Bahamas. *Journal of Crustacean Biology* 16: 754–762.

Suzuki, H., and C. L. McLay. 1998. Gill-cleaning mechanisms of *Paratya curvirostris* (Caridea, Atyidae) and comparisons with seven species of Japanese atyid shrimps. *Journal of Crustacean Biology* 18: 253–270.

Svavarsson, J. 1984. Description of the male of *Pseudomesus brevicornis* Hansen, 1916 (Isopoda, Asellota, Desmosomatidae) and rejection of the family Pseudomesidae. *Sarsia* 69:37–44.

———. 1987. Reevaluation of *Katianira* in Arctic waters and erection of a new family, Katianiridae (Isopoda: Asellota). *Journal of Crustacean Biology* 7:704–720.

Swanson, K. M. 1989a. Ostracod phylogeny and evolution—a manawan perspective. *Courier Forschungsinstitut Senckenberg* 113:11–20.

———. 1989b. *Manawa staceyi* n. sp. (Punciidae, Ostracoda): soft anatomy and ontongeny. *Courier Forschungsinstitut Senckenberg* 113:235–249.

———. 1990. The punciid ostracod—a new crustacean evolutionary window. *Courier Forschungsinstitut Senckenberg* 123:11–18.

———. 1991. Distribution, affinities and origin of the Punciidae (Crustacea: Ostracoda). In *Proceedings of the 1990 International Crustacean Conference,* ed. P. J. F. Davie and R. H. Quinn. *Memoirs of the Queensland Museum* 31:77–92.

Tabacaru, I., and D. L. Danielopol. 1996a. Phylogénie des Isopodes terrestres. *Comptes Rendus Hebdomadaires des Seances de l'Academie des Sciences* 319:71–80.

———. 1996b. Phylogenèse et convergence chez les Isopodes terrestres. *Vie et Milieu* 46:171–181.

Tabuki, R., and T. Hanai. 1999. A new sigillid ostracod from submarine caves of the Ryukyu Islands, Japan. *Paleontology* 42:569–593.

Takeuchi, I. 1993. Is the Caprellidea a monophyletic group? *Journal of Natural History* 27:947–964.

Tasch, P. 1969. Branchiopoda. In *Treatise on invertebrate paleontology, part R, Arthropoda 4,* ed. R. C. Moore, R128–R191. Lawrence, Kansas: The Geological Society of America and the University of Kansas Press.

Tavares, M. S. 1991. Espèces nouvelles de Cyclodorippoidea Ortmann et remarques sur les genres *Tymolus* Stimpson et *Cyclodorippe* A. Milne Edwards (Crustacea, Decapoda, Brachyura). *Bulletin de Museum National d'Histore Naturelle,* series 4, 12:623–648.

———. 1993. Crustacea Decapoda: Les Cyclodorippidae et Cymonomidae de l'Indo-Ouest-Pacifique à l'exclusion du genre *Cymonomus.* In *Résultats des Campagnes MUSORSTOM,* vol. 10, ed. A. Crosnier. *Mémoires du Muséum National d'Histoire Naturelle,* vol. 156, 253–313.

———. 1998. Phyllotymolinidae, nouvelle famille de Brachyoures Podotremata (Crustacea, Decapoda). *Zoosystema* 20:109–122.

Taylor, D. J., T. J. Crease, and W. M. Brown. 1999. Phylogenetic evidence for a single long-lived clade of crustacean cyclic parthenogens and its implications for the evolution of sex. *Proceedings of the Royal Society of London* 266B:791–797.

Taylor, R. S., and F. R. Schram. 1999. Meiura (anomalan and brachyuran crabs). In *Functional morphology of the invertebrate skeleton,* ed. E. Savazzi, 517–528. Chichester: John Wiley.

Taylor, R. S., F. R. Schram, and Y. Shen. 1999. A new crayfish family (Decapoda: Astacida) from the Upper Jurassic of China, with a reinterpretation of other Chinese crayfish taxa. *Paleontological Research* 3: 121–136.

Taylor, R. S., Y. Shen, and F. R. Schram. 1998. New pygocephalomorph crustaceans from the Permian of China and their phylogenetic relationships. *Paleontology* 41:815–834.

Tchindonova, J. G. 1981. New data on the systematic position of some deep-sea mysids (Mysidacea, Crustacea) and their distribution in the world ocean. In *Biology of the Pacific Ocean depths,* ed. N. G. Vinogradova, 24–33. Vladivostok: Academy of Sciences of the USSR. [In Russian]

Telford, M. J., and R. H. Thomas. 1995. Demise of the Atelocerata? *Nature* 376:123–124.

Thatcher, V. E. 1986. The parasitic crustaceans of fishes from the Brazilian Amazon, 16, *Amazonicopeus elongatus* gen. et sp. nov. (Copepoda: Poecilostomatoida) with the proposal of Amazonicopeidae fam. nov. and remarks on its pathogenicity. *Amazoniana* 10:49–56.

Thatcher, V. E., and B. A. Robertson. 1984. The parasitic crustaceans of fishes from the Brazilian Amazon. 11. Vaigamidae fam. nov. (Copepoda: Poecilostomatoida) with males and females of *Vaigamus retrobarbatus* gen. et sp. nov. and *V. spinicephalus* sp. nov. from plankton. *Canadian Journal of Zoology* 62: 716–729.

Thiéry, A. 1996. Branchiopodes I. Ordres des Anostracés, Notostracés, Spinicaudata et Laevicaudata (Anostraca Sars, 1867—Notostraca Sars, 1867—Spinicaudata Linder, 1945—Laevicaudata Linder, 1945). In *Traité de Zoologie. Anatomie, Systematique, Biologie. Tome VII, Fascicule II. Généralités (suite) et Systématique,* ed. J. Forest, 287–351. Paris: Masson, 1002 pp.

Thurston, M. H. 1982. *Cheus annae,* new genus, new species (Cheidae, new family), a fossorial amphipod from the Falkland Islands. *Journal of Crustacean Biology* 2:410–419.

———. 1989. A new species of *Valettia* (Crustacea: Amphipoda) and the relationship of the Valettidae to the

Lysianassoidea. *Journal of Natural History* 23: 1093–1107.

Trilles, J.-P. 1999. Ordre des Isopodes sous-ordre des Épicarides (Epicaridea Latreille, 1825). In *Traité de Zoologie. Anatomie, Systématique, Biologie. Tome VII, Fascicule IIIA. Crustacés Péracarides,* ed. J. Forest. Memoires de l'Institut Oceanographique Fondation Albert Ier, Prince de Monaco, No. 19, 279–352.

Tshudy, D., and L. E. Babcock. 1997. Morphology-based phylogenetic analysis of the clawed lobsters (family Nephropidae and the new family Chilenophoberidae). *Journal of Crustacean Biology* 17:253–263.

Tucker, A. B. 1998. Systematics of the Raninidae (Crustacea: Decapoda: Brachyura) with accounts of three new genera and two new species. *Proceedings of the Biological Society of Washington* 111:320–371.

Tudge, C. C. 1991. Spermatophore diversity within and among the hermit crab families, Coenobitidae, Diogenidae and Paguridae (Paguroidea, Anomura, Decapoda). *Biological Bulletin* 181:238–247.

———. 1992. Comparative ultrastructure of hermit crab spermatozoa (Paguroidea, Anomura, Decapoda). *Journal of Crustacean Biology* 12:397–409.

———. 1995. Ultrastructure and phylogeny of the spermatozoa of the infraorders Thalassinidea and Anomura (Decapoda, Crustacea). In *Advances in spermatozoal phylogeny and taxonomy,* ed. B. G. M. Jamieson, J. Ausio, and J.-L. Justine. *Mémoires du Muséum National d'Histoire Naturelle,* vol. 166, 251–263.

———. 1997a. Spermatological evidence supports the taxonomic placement of the Australian endemic hairy stone crab, *Lomis hirta* (Decapoda, Anomura, Lomidae). *Memoirs of the Museum of Victoria* 56: 235–244.

———. 1997b. Phylogeny of the Anomura (Decapoda, Crustacea): spermatozoan and spermatophore morphological evidence. *Contributions to Zoology* 67: 125–141.

———. 1999a. Ultrastructure of the spermatophore lateral ridge in hermit crabs (Decapoda, Anomura, Paguroidea). *Crustaceana* 72:77–84.

———. 1999b. Spermatophore morphology in the hermit crab families Paguridae and Parapaguridae (Paguroidea, Anomura, Decapoda). *Invertebrate Reproduction and Development* 35:203–214.

Tudge, C. C., B. G. M. Jamieson, L. Sandberg, and C. Erseus. 1998a. Ultrastructure of the mature spermatozoon of the king crab *Lithodes maja* (Lithodidae, Anomura, Decapoda): further confirmation of a lithodid-pagurid relationship. *Invertebrate Biology* 117:57–66.

Tudge, C. C., B. G. M. Jamieson, M. Segonzac, and D. Guinot. 1998b. Spermatozoal ultrastructure in three species of hydrothermal vent crab, in the genera *Bythograea, Austinograea* and *Segonzacia* (Decapoda, Brachyura, Bythograeidae). *Invertebrate Reproduction and Development* 34:13–23.

Tudge, C. C., G. C. B. Poore, and R. Lemaitre. 2000. Preliminary phylogenetic analysis of generic relationships within the Callianassidae and Ctenochelidae (Decapoda: Thalassinidea: Callianassoidea). *Journal of Crustacean Biology* 20(special no. 2):129–149.

Tudge, C. C., D. M. Scheltinga, and B. G. M. Jamieson. 1999. Spermatozoal ultrastructure in the Hippoidea (Anomura, Decapoda). *Journal of Submicroscopic Cytology and Pathology* 31:1–13.

Turbeville, J. M., D. M. Pfeiffer, K. G. Field, and R. A. Raff. 1991. The phylogenetic status of arthropods, as inferred from 18S rRNA sequences. *Molecular Biology and Evolution* 8:669–686.

Vader, W. 1972. A list of amphipod genera and species described by W. Lilljeborg. *Amphipod Newsletter* 2: 13–15.

Vader, W., A. Baldinger, and T. Krapp-Schickel. 1998. IXth International Amphipod Meeting (Kronenberg, Germany). *Ecdysiast* (Newsletter of The Crustacean Society,) vol. 17, no. 2:6–7.

Van Dover, C. L., R. H. Gore, and P. Castro. 1986. *Echinoecus pentagonus* (A. Milne Edwards, 1879): larval development and systematic position (Crustacea: Brachyura: Xanthoidea nec Parthenopoidea). *Journal of Crustacean Biology* 6:757–776.

Van Lieshout, S. E. N. 1983. Amsterdam expedition to the West Indian Islands, report 27. Calabazoidea, new suborder of stygobiont Isopoda, discovered in Venezuela. *Bijdragen tot de Dierkunde* 53:165–177.

Vannier, J., and K. Abe. 1995. Size, body plan, and respiration in the Ostracoda. *Paleontology* 38:843–873.

Vannier, J., M. Williams, and D. J. Siveter. 1997. The Cambrian origin of the circulatory system of crustaceans. *Lethaia* 30:169–184.

Vereshchaka, A. L. 1996. A new genus and species of caridean shrimp (Crustacea: Decapoda: Alvinocarididae) from North Atlantic hydrothermal vents. *Journal of the Marine Biological Association of the United Kingdom* 76:951–961.

———. 1997a. A new family for a deep-sea caridean shrimp from North Atlantic hydrothermal vents. *Journal of the Marine Biological Association of the United Kingdom* 77:425–438.

———. 1997b. A new family and superfamily for a deep-sea caridean shrimp from the *Galathea* collections. *Journal of Crustacean Biology* 17:361–373.

Vinogradov, M. E., A. F. Volkov, and T. N. Semenova. 1982 (1996). *Hyperiid amphipods (Amphipoda, Hyperiidea) of the world oceans.* Editor of 1996 English version: D. Siegel-Causey, translated under agreement with the Smithsonian Institution Libraries, Washington, D.C., by Amerind Publishing Co. Pvt. Ltd., New Delhi. [English translation of Amfipodygiperiidy (Amphipoda, Hyperiidea) mirovogo okeana, Leningrad: Nauka Publishers.]

Vonk, R., and F. R. Schram. 1998. On the distribution and phylogeny of ingolfiellid amphipods. *Proceedings and Abstracts of the 4th International Crustacean Congress, Amsterdam*: 58 (abstract 56).

Wägele, J.-W. 1983. On the origin of the Microcerberidae (Crustacea: Isopoda). *Zeitschrift für Zoologische Systematik und Evolutionsforschung* 21:249–262.

———. 1989. Evolution und phylogenetisches System der Isopoda. Stand der Forschung und neue Erkenntnisse. *Zoologica (Stuttgart)* 140:1–262.

———. 1994. Review of methodological problems of 'computer cladistics' exemplified with a case study on isopod phylogeny (Crustacea: Isopoda). *Zeitschift für Zoologische Systematik und Evolutionsforschung* 32:81–107.

———. 1996. The theory and methodology of phylogenetic systematics is still evolving: a reply to Wilson. *Vie et Milieu* 46:183–184.

Wägele, J.-W., and A. Brandt. 1988. *Protognathia* n. gen. *bathypelagica* (Schultz, 1977) rediscovered in the

Weddell Sea: a missing link between the Gnathiidae and the Cirolanidae. *Polar Biology* 8:359–365.

Wägele, J.-W., and G. Stanjek. 1995. Arthropod phylogeny inferred from partial 12S rRNA revisited: monophyly of the Tracheata depends on sequence alignment. *Journal of Zoological Systematics and Evolutionary Research* 33:75–80.

Wagner, H. P. 1994. A monographic review of the Thermosbaenacea. *Zoologische Verhandelingen* 291:1–338.

Walker-Smith, G. K. 2000. *Levinebalia maria,* a new genus and species of Leptostraca (Crustacea) from Australia. *Memoirs of Museum Victoria* 58:137–148.

Walker-Smith, G. K., and G. C. B. Poore. 2001. A phylogeny of the Leptostraca (Crustacea) with keys to families and genera. *Memoirs of Museum Victoria* 58:383–410.

Walossek, D. 1993. The Upper Cambrian *Rehbachiella kinnekullensis* and the phylogeny of Branchiopoda and Crustacea. *Fossils and Strata* 32:1–202.

———. 1995. The Upper Cambrian *Rehbachiella,* its larval development, morphology, and significance for the phylogeny of Branchiopoda and Crustacea. *Hydrobiologia* 298:1–13.

———. 1999. On the Cambrian diversity of Crustacea. In *Crustaceans and the biodiversity crisis. Proceedings of the 4th International Crustacean Congress, Amsterdam,* vol. 1, ed. F. R. Schram and J. C. von Vaupel Klein, 3–27. Leiden: Brill.

Walossek, D., and K. J. Müller. 1990. Stem-lineage crustaceans from the Upper Cambrian of Sweden and their bearing upon the position of *Agnostus. Lethaia* 23:409–427.

———. 1992. The 'Alum Shale Window'—contribution of 'Orsten' arthropods to the phylogeny of Crustacea. *Acta Zoologica* 73:305–312.

———. 1994. Pentastomid parasites from the Lower Paleozoic of Sweden. *Transactions of the Royal Society of Edinburgh, Earth Sciences* 85:1–37.

———. 1997. Cambrian "Orsten"-type arthropods and the phylogeny of Crustacea. In *Arthropod relationships,* ed. R. A. Fortey, 139–153. Systematics Association Special Volume Series 55. London: Chapman and Hall.

———. 1998. Early arthropod phylogeny in light of the Cambrian "Orsten" fossils. In *Arthropod fossils and phylogeny,* ed. G. D. Edgecombe, 185–231. New York: Columbia University Press.

Walossek, D., J. E. Repetski, and K. J. Müller. 1994. An exceptionally preserved parasitic arthropod, *Heymoniscambria taylori* n. sp. (Arthropoda incertae sedis: Pentastomida), from Cambrian-Ordovician boundary beds of Newfoundland, Canada. *Canadian Journal of Earth Sciences* 31:1664–1671.

Walossek, D., and H. Szaniawski. 1991. *Cambrocaris baltica,* n. gen. n. sp., a possible stem-lineage crustacean from the Upper Cambrian of Poland. *Lethaia* 24:363–378.

Watling, L. 1981. An alternative phylogeny of peracarid crustaceans. *Journal of Crustacean Biology* 1:201–210.

———. 1983. Peracaridan disunity and its bearing on Eumalacostracan phylogeny with a redefinition of Eumalacostracan superorders. In *Crustacean issues 1. Crustacean phylogeny,* ed. F. R. Schram, 213–228. Rotterdam: A. A. Balkema Press, 372 pp.

———. 1998. On peracarid unity: who goes, who stays? *Proceedings and Abstracts of the 4th International Crustacean Congress, Amsterdam*: 71 (abstract 518).

———. 1999a. The place of the Hoplocarida in the malacostracan pantheon. *Program and Abstracts, The Crustacean Society 1999 Summer Meeting, Lafayette, Louisiana*: 47.

———. 1999b. Toward understanding the relationships of the peracaridan orders: the necessity of determining exact homologies. In *Crustaceans and the biodiversity crisis. Proceedings of the 4th International Crustacean Congress, Amsterdam,* vol. 1, ed. F. R. Schram and J. C. von Vaupel Klein, 73–89. Leiden: Brill.

Watling, L., C. H. J. Hof, and F. R. Schram. 2000. The place of the Hoplocarida in the malacostracan pantheon. *Journal of Crustacean Biology* 20(special number 2):1–11.

Watling, L., and I. Kornfield. 1996. Cumacean web page, URL http://nature.umesci.maine.edu/pub/cumacea.html.

Whatley, R. C., D. J. Siveter, and I. D. Boomer. 1993. Arthropoda (Crustacea: Ostracoda). In *The fossil record 2,* ed. M. J. Benton, 343–356. London: Chapman and Hall.

Wheeler, W. C. 1997. Sampling, groundplans, total evidence and the systematics of arthropods. In *Arthropod relationships,* ed. R. A. Fortey and R. H. Thomas, 87–96. Systematics Association Special Volume Series 55. London: Chapman and Hall.

———. 1998. Molecular systematics and arthropods. In *Arthropod fossils and phylogeny,* ed. G. D. Edgecombe, 9–32. New York: Columbia University Press.

Wheeler, W. C., P. Cartwright, and C. Y. Hayashi. 1993. Arthropod phylogeny: a combined approach. *Cladistics* 9:1–39.

Willen, E. 1999. Preliminary revision of the Pseudotachidiidae Lang, 1936 (Copepoda, Harpacticoida). *Courier Forschungsinstitut* 215:221–225.

Williams, A. B. 1979. A new crab family from shallow waters of the West Indies (Crustacea: Decapoda: Brachyura). *Proceedings of the Biological Society of Washington* 92:399–413.

———. 1980. A new crab family from the vicinity of submarine thermal vents on the Galapagos Rift (Crustacea: Decapoda: Brachyura). *Proceedings of the Biological Society of Washington* 93:443–472.

———. 1988. *Lobsters of the world—an illustrated guide.* Huntington, New York: Osprey Books, 186 pp.

Williams, T. A., and L. M. Nagy. 1995. Brine shrimp add salt to the stew. *Current Biology* 5:1330–1333.

Williams, W. D., and J. L. Barnard. 1988. The taxonomy of crangonyctoid Amphipoda (Crustacea) from Australian fresh waters: foundation studies. *Records of the Australian Museum,* 10(supplement):1–180.

Williamson, D. I. 1976. Larval characters and the origin of crabs (Crustacea, Decapoda, Brachyura). *Thalassia Jugoslavica* 10:401–414.

———. 1982. Larval morphology and diversity. In *The biology of Crustacea,* vol. 2. *Embryology, morphology, and genetics,* ed. L. G. Abele, 43–110. New York: Academic Press.

———. 1988a. Evolutionary trends in larval form. In *Aspects of decapod crustacean biology,* ed. A. A. Fincham and P. Rainbow, 11–25. London: Symposium of the Zoological Society of London 59, 375 pp.

———. 1988b. Incongruous larvae and the origin of some invertebrate life histories. *Progress in Oceanography* 19:87–116.

Wills, M. A. 1997. A phylogeny of recent and fossil Crustacea derived from morphological characters. In *Arthropod relationships,* ed. R. A. Fortey and R. H. Thomas, 189–209. Systematics Association Special Volume Series 55. London: Chapman and Hall.

———. 1998. Crustacean disparity through the Phanerozoic: comparing morphological and stratigraphic data. *Biological Journal of the Linnean Society* 65: 455–500.

Wills, M. A, D. E. Briggs, R. A. Fortey, and M. Wilkinson. 1995. The significance of fossils in understanding arthropod evolution. *Verhandlungen der Deutschen Zoologischen Gesellschaft* 88:203–215.

Wills, M. A., D. E. G. Briggs, R. A. Fortey, M. Wilkinson, and P. H. A. Sneath. 1998. An arthropod phylogeny based on fossil and Recent taxa. In *Arthropod fossils and phylogeny,* ed. G. D. Edgecombe, 33–105. New York: Columbia University Press.

Wilson, G. D. F. 1986. Pseudojaniridae (Crustacea: Isopoda), a new family for *Pseudojanira stenetrioides* Barnard, 1925, a species intermediate between the asellote superfamilies Stenetrioidea and Janiroidea. *Proceedings of the Biological Society of Washington* 99:350–358.

———. 1987. The road to the Janiroidea: comparative morphology and evolution of the asellote isopod crustaceans. *Zeitschrift für Zoologische Systematik und Evolutionsforschung* 25:257–280.

———. 1989. A systematic revision of the deep-sea subfamily Lipomerinae, of the isopod crustacean family Munnopsidae. *Bulletin of the Scripps Institution of Oceanography* (University of California, San Diego) 27:1–138.

———. 1992. Computerized analysis of crustacean relationships. *Acta Zoologica* 73:383–389.

———. 1994. A phylogenetic analysis of the isopod family Janiridae (Crustacea). *Invertebrate Taxonomy* 8: 749–766.

———. 1996. Of uropods and isopod crustacean trees: a comparison of "groundpattern" and cladistic methods. *Viet et Milieu* 46:139–153.

Wilson, G. D. F., and R. T. Johnson. 1999. Ancient endemism among freshwater isopods (Crustacea, Phreatoicidea). In *The other 99%. The conservation and biodiversity of invertebrates,* ed. W. Ponder and D. Lunney, 264–268. Mosman, New South Wales: Transactions of the Royal Zoological Society of New South Wales.

Wilson, G. D. F., and S. J. Keable. 1999. A new genus of phreatoicidean isopod (Crustacea) from the north Kimberly region, western Australia. *Zoological Journal of the Linnean Society* 126:51–79.

———. 2001. Systematics of the Phreatoicidea. In *Crustacean Issues 13. Isopod systematics and evolution,* ed. B. Kensley and R. C. Brusca, 175–194. Rotterdam: A. A. Balkema Press, 357 pp.

Wilson, K., V. Cahill, E. Ballment, and J. Benzie. 2000. The complete sequence of the mitochondrial genome of the crustacean *Penaeus monodon*: are malacostracan crustaceans more closely related to insects than to branchiopods? *Molecular Biology and Evolution* 17:863–874.

Wingstrand, K. G. 1972. Comparative spermatology of a pentastomid *Raeillietiella hemidactyli* and a branchiuran crustacean *Argulus foliaceus* with a discussion of pentastomid relationships. *Kongelige Danske Videnskabarens Selskab Biologiske Skrifter* 19:1–72.

———. 1978. Comparative spermatology of the Crustacea Entomostraca. 1. Subclass Branchiopoda. *Kongelige Danske Videnskabarens Selskab Biologiske Skrifter* 22:1–66.

———. 1988. Comparative spermatology of the Crustacea Entomostraca. 2. Subclass Ostracoda. *Kongelige Danske Videnskabarens Selskab Biologiske Skrifter* 32:1–149.

Winnepenninckx, B. M. H., T. Backeljau, and R. M. Kristensen. 1998. Relations of the new phylum Cycliophora. *Nature* 393:636–638.

Wolff, T. 1989. The genera of Santiidae Kussakin, 1988, with the description of a new genus and species (Crustacea, Isopoda, Asellota). *Steenstrupia* 15:177–191.

Wray, G. A., J. S. Levinton, and L. H. Shapiro. 1996. Molecular evidence for deep Precambrian divergences among metazoan phyla. *Science* 274:568–573.

Yager, J. 1981. Remipedia, a new class of crustaceans from a marine cave in the Bahamas. *Journal of Crustacean Biology* 1:328–333.

———. 1989a. The male reproductive system, sperm, and spermatophores of the primitive hermaphroditic, remipede crustacean *Speleonectes benjamini. Invertebrate Reproduction and Development* 15:75–81.

———. 1989b. *Pleomothra apletocheles* and *Godzilliognomus frondosus,* two new genera and species of remipede crustaceans (Godzilliidae) from anchialine caves in the Bahamas. *Bulletin of Marine Science* 44: 1195–1206.

———. 1991. The Remipedia (Crustacea): recent investigations of their biology and phylogeny. *Verhandlungen der Deutschen Zoologischen Gesellschaft, Stuttgart* 84:261–269.

Yager, J., and J. H. Carpenter. 1999. *Speleonectes epilimnius* new species (Remipedia, Speleonectidae) from surface water of an anchialine cave on San Salvador Island, Bahamas. *Crustaceana* 72:965–977.

Yager, J., and W. F. Humphreys. 1996. *Lasionectes exleyi,* sp. nov., the first remipede crustacean recorded from Australia and the Indian Ocean, with a key to the world species. *Invertebrate Taxonomy* 10:171–187.

Yager, J., and F. R. Schram. 1986. *Lasionectes entrichoma,* new genus, new species (Crustacea: Remipedia) from anchialine caves in the Turks and Caicos, British West Indies. *Proceedings of the Biological Society of Washington* 99:65–70.

Yamaguchi, T., and W. A. Newman. 1990. A new and primitive barnacle (Cirripedia: Balanomorpha) from the North Fiji Basin abyssal hydrothermal field, and its evolutionary implications. *Pacific Science* 44:135–155.

Yamaguti, S. 1963. *Parasitic Copepoda and Branchiura of fishes.* New York: Interscience Publishers (John Wiley & Sons), i–vii + 1104 pp.

Young, P. S. (editor). 1998. *Catalogue of Crustacea of Brazil.* Rio de Janeiro: Museo Nacional, Série Livros 6, i–xvii + 717 pp.

Zhang, W., Y. Shen, and N. Shaowu. 1990. Discovery of Jurassic conchostracans with well preserved soft parts and notes on its biological significance. *Palaeontologia Cathayana* 5:311–352.

# APPENDIX I. COMMENTS AND OPINIONS

The following comments and opinions were provided by colleagues (all of whom are listed in Appendix II) after seeing the penultimate draft of the classification. The authors wish to gratefully acknowledge them for allowing us to reproduce their remarks. References are listed after each comment only if those references are not already listed in our Literature Cited section. Some authors did not supply full references; consequently, references may be missing for some papers cited below.

## CRUSTACEA (GENERAL)

The authors choose to treat the Crustacea as a monophyletic group and thus find it justifiable to produce an updated classification for organizing museum collections and helping students of crustaceans to search unfamiliar taxa. It should thus become a useful taxonomic tool. I find much merit in (1) the exposition of reasons for preferred arrangements and (2) the attempt to introduce readers to alternative opinions. The permanent drawback of this compilation (considered by the authors) is that taxa are not justified by diagnostic characters.

As a means of reflecting some current phylogenetic ideas on crustaceans, however, the present attempt will be considered obsolete almost immediately by some workers. The monophyly of the Crustacea is far from settled. In fact, in my opinion, it is very unlikely. The mandibulate arthropods are traditionally divided into two grades (crustaceans and tracheates), and it is obvious that the closest relatives of the terrestrial tracheates should be sought among aquatic crustaceans. If this scenario is reasonable, the Crustacea become, in principle, a nonmonophyletic grade-group. The Remipedia and Malacostraca have been pinpointed as two successive outgroups of the Tracheata (Moura and Christoffersen, 1996). If there is merit in such a proposal, an incorrect assumption of monophyly could immediately account for many discrepancies noted among cladistic papers establishing the position and internal relationships of the Crustacea. Researchers striving for a phylogenetic arrangement of the crustaceans should not exclude the terrestrial descendants of crustaceans from their system. For these reasons, rather than a practical, largely consensual, and authority-based classification of the Recent Crustacea, we need to reconstruct the system of the Mandibulata (apparently the smallest clade that includes all the so-called crustaceans, as well as their myriapod and hexapod descendants). Furthermore, apomorphic characters need to be provided to distinguish acceptable monophyletic taxa from unstudied, unknown, or unresolved traditional taxa. Let me suggest that this become another demanding, but long overdue, story.

*Submitted by Martin L. Christoffersen, Federal University of Paraíba, Brazil*

## BRANCHIOPODA AS PRIMITIVE

In regards to your first argument here, there are three different sets of authors who cannot confirm a branchiopod affinity for this taxon [*Rehbachiella*] and consequently there in fact *may* be no Cambrian branchiopods. The second part of your argument, that there are neither Cambrian cephalocarids, nor remipedes, is a non-sequitor. The late Ralph Gordon Johnson used to say about the apparent age of fossils "Things are always older than you think they are." An example of which relates to those Carboniferous remipedes; there is in fact something in the Silurian of Wisconsin, yet undescribed, that may be a remipede. So, your first argument is weak.

Your second argument, derived from apomorphic development, would seem to be valid, at least under traditional assumptions. However, two points might be mentioned in this regard. The weakest point relates to the basic assumption of anamorphy = primitive. Certain aspects emerging from developmental genetics might suggest an alternative; however, this needs to be developed and published (something I have not had time to do as yet). Nevertheless, if we consider the matter in strictly cladistic terms, if as you correctly state that anamorphy is unique to branchiopods, within Crustacea sensu stricto the issue of plesiomorphy is not resolved—branchiopods have it, but non-branchiopods (apparently) don't. If you add outgroups from the "other Mandibulata," in an attempt to polarize patterns of development, then if insects are in fact a sister group to crustaceans, epimorphy could be argued as plesiomorphic.

Third, the molecular data cited here is not being employed properly by you. The distinctness of branchiopods here in the papers you cite is stronger than you indicate. For example, Spears and Abele (1997) under certain assumptions actually pull branchiopods into the hexapods, which possibly indicates crustacean polyphyly. Of course you say that branchiopods are (might be) closer to other groups of arthropods—a fair judgment. If true, that would indicate that the position of branchiopods far exceeds that of a potential "basal group" of crustaceans. Primitiveness under those circumstances has nothing to do with it.

In short, you are wise not to create any additional taxonomic categories. Moreover, your three-pronged argument would appear to be not clearly drawn at all.

On the ancestral crustacean . . . you remark that Schram and Hof (1998) obtain a clade Phyllopoda. First, if you look at the paper carefully, we sometimes get a phyllopodan clade, and sometimes not—depending on the assumptions and inclusiveness of the database employed. Contrary to Schram (1986), I think Hof and I would state that the issue of whether or not there is a monophyletic clade Phyllopoda is indeed an open one—which is not what your sentence says. Second, just because one clade in a comprehensive analysis does not find wide favor does not necessarily call other aspects of the analysis into question. [*Editors'note: In our penultimate draft, we criticized the recognition of the Phyllopoda by Schram and Hof, and then used that criticism to cast doubt on other of their findings in that paper; this unfair criticism has since been removed.*] What the main conclusion of Schram and Hof indicated was that the issue of crustacean phylogenetic relationships has more mileage in it before we hope to approach a solution. That ought to be conveyed in your text at this point.

*Submitted by Frederick R. Schram,*
*Zoölogisches Museum, Amsterdam*

## BRANCHIOPODA AS PRIMITIVE

You state several places that you place the Branchiopoda as the sister group to the remaining crustaceans. This may be correct, but you mention no arguments. The only possible arguments could be characters shared by the remaining crustaceans that would set the Branchiopoda aside. It is not enough to state that they [look] very primitive and that some of them look like some of the 'Orsten' fossils. I agree, of course, that the branchiopods ARE indeed some of the most primitive Recent Crustacea we have, but this doesn't automatically give them sister group position to the rest (only synapomorphies for the remaining . . . , as mentioned above). It is NOT difficult to imagine the branchiopods (or the cephalocarids) placed a little bit up in the system. It would only require that the primitive features that they have are retained a couple of nodes, and that those that actually are branched off first (Malacostraca, Remipedia, whatever) have attained their special modifications independently from other Crustacea.

So, to summarize, the discussion of which Crustacea is the most primitive to look at, and which is the sister group to the rest, is a mixture of two discussions which actually should be separate. The two discussions have been treated as one when certain other authors have been discussing the same for cephalocarids and remipedes, I know, but it does not make the discussion more sensible. I believe plenty of examples could be mentioned where the sister group to a larger group is far from being the best candidate as the most primitive one. To take an example from animals we are both interested in: If for example notostracans are the sister group to all the 'bivalved' branchiopods, it doesn't follow that they also are the most primitive. This is the same story for the possible sister group to the Crustacea. We should not exclude any of the derived forms from having that honorary position. Only synapomorphies uniting the rest can place a taxon in this position.

My advice would be to skip the idea of branchiopod as sister group to the rest, unless you provide arguments. But of course, you should retain the point of branchiopods being quite primitive (based on similarities to certain 'Orsten' fossils), but I think it is impossible and subjective to distinguish between the branchiopods and the cephalocarids in this respect. [Both] look like certain 'Orsten' fossils, and not least the cephalocarids. There is not [an] objective way to say which is most primitive, because it depends on the feature you focus on. So, perhaps you should mention both taxa as the best candidates to being 'primitive'.

Also, I simply don't understand how you can say that we 'are treating the class Branchiopoda as the most primitive of the Crustacea' when this is not included in your classification. It sounds like you don't believe it enough to actually include it (by finding a name for the rest). In my opinion, it contains no information about primitivity to mention it as the first of the classes in your classification.

*Submitted by Jørgen Olesen,*
*University of Copenhagen, Denmark*

## BRANCHIOPODA

Elucidation of the relationships of the "cladoceran" and "conchostraca" branchiopods appears to have reached what is doubtless a temporary impasse. Morphology seems to be saying one thing, some molecular evidence another. In morphology, the "cladoceran" orders differ much from each other, and attempts to unite them are unsatisfactory. On his own estimation, Olesen (1998), who would do so, feels that the monophyly of the "Cladocera" "may not seem well supported" by his cladistic analysis. In fact, of five characters used in support, three are wrong, one is of no significance, and the other is but a small, to be expected, adaptive change that could have happened more than once. The four constituent groups, which merit ordinal rank, differ from each other more than do the various orders of the Copepoda. Although some copepods are modified for parasitic habits, some representatives of all orders retain various fundamental similarities.

Olesen himself says that his analysis does not support the "Conchostraca," nor, incidentally, the Spinicaudata, a well-defined component of that group, especially if the divergent *Cyclestheria* is segregated from it. Nevertheless, he unites the morphologically diverse "cladoceran" orders with the unsupported, and very different, "Conchostraca" as the "Diplostraca," which compounds the difficulties. All the alleged synapomorphies of the "Diplostraca" are incorrect (Fryer, 1999b). Walossek's (1993, 1995) less detailed attempt to demonstrate the same relationship fails for similar reasons.

The Spinicaudata was fully differentiated at least as long ago as the early Devonian. Ephippia of even extant genera of the "cladoceran" order Anomopoda are known from the Lower Cretaceous, and molecular evidence suggests that *Daphnia* originated more than 200 My ago (Colbourne and Hebert,

1996). The order must be extremely ancient. If the "cladoceran" orders prove to be monophyletic, they must be of extremely ancient origin. The most convincing molecular evidence of affinity of the "cladoceran" orders is that in all four the V4 and V7 regions of the small subunit ribosomal RNA possesses four helices, three of which are present in *Cyclestheria* but are otherwise so far unique (Crease and Taylor, 1998). *Cyclestheria,* long regarded as a somewhat recalcitrant spinicaudatan, has often been cast in the role of ancestor of the "Cladocera"—without however demonstrating how such different orders as the Anomopoda and Haplopoda could have been derived from it. Although the helices are very different in length and primary sequences of their distal ends in the different orders, their locations, secondary structures, and primary sequences at their proximal ends are conserved, which suggests homology. None of these peculiarities is shared with the Spinicaudata, within which order *Cyclestheria* was long included and to which it is vastly more similar in morphology than it is to any "cladoceran" order! According to some investigators, evidence deduced from 18S ribosomal DNA supports these relationships (Spears and Abele, 2000). However, according to Dumont (2000), "ongoing molecular work using the full sequence of the 18S rDNA nuclear gene" not only confirms the distinction of that order "but also suggests that the Onychopoda might even be more closely related to the Anostraca than with the cladoceran orders Ctenopoda and Anomopoda." Note, also, that the widely accepted 18S rRNA phylogenetic tree of the Protozoa has now been seriously questioned, and is probably unreliable (Phillippe and Adoutte, 1998)!

With qualifications, some molecular evidence is seductive and welcome, but is contradicted by other molecular findings, and cannot gainsay either the great morphological differences between the groups concerned, or the failure to justify either the "Cladocera," "Conchostraca," or "Diplostraca" by cladistic analyses. To change the classification of these animals on the basis of still-contentious molecular evidence while ignoring the larger corpus of information now accumulated, not only on morphology but on morphology whose functional significance is sometimes understood, and on life histories, would merely upset what may indeed eventually prove to be only an interim scheme, but one which for the time being is perfectly serviceable. As Avise (1994) notes, morphological and molecular evolution may proceed at different rates, and the overall magnitude of genetic distance between taxa is not necessarily the only, or the best, guide to phylogenetic relationships within groups.

The subclasses Sarsostraca and Phyllopoda seem to be unnecessary. The latter name has also already been a source of much confusion. A case can be made for the Notostraca as being as distinctive as the Anostraca, which alone renders grouping into subclasses untenable.

### Additional References


Avise, J. C. 1994. Molecular markers, natural history and evolution. New York: Chapman and Hall.

Colbourne, J. K., and P. D. N. Hebert. 1996. The systematics of the North American *Daphnia* (Crustacea: Anomopoda): a molecular phylogenetic approach. *Philosophical Transactions of the Royal Society of London* 351B:349–360.

Dumont, H. J. 2000. Endemism in the Ponto-Caspian fauna, with special emphasis on the Onychopoda (Crustacea). *Advances in Ecological Research* 31:181–196.

Phillippe, H., and A. Adoutte. 1998. The molecular phylogeny of Eukaryota: solid facts and uncertainties. In *Evolutionary relationships among Protozoa,* eds. G. H. Coombs et al., 25–56. London: Chapman and Hall.


*Submitted by Geoffrey Fryer,*
*University of Lancaster, United Kingdom*

## BRANCHIOPODA

I am not sure that you should not include the Ilyocryptidae in your classification. After all, it is a quite serious action not to follow the advice of the most important Recent taxonomist working in the Cladocera that we have (N. N. Smirnov). Especially since you follow so many other taxonomists in their suggestions. You present no arguments for not doing so. One could argue that an eventual splitting of the Macrothricidae should await a phylogenetic revision, but such a revision is likely not to appear in due time. It is true that the change suggested by Smirnov may not be based on phylogenetic criteria (and the remaining macrothricids may still be paraphyletic), but the same could be said about so much of your classification anyway, as you mention a couple of times.

I think when it comes to the lower level classification, I believe it would be wise to follow the advice of the people actually working on the taxa, unless you have personal, strong arguments no to do so. The case of the 'Moinidae' is different because Fryer convincingly argues for their unity with the rest of the Daphniidae. You could also cite his 1991 monograph on Daphniidae adaptive radiation here.

The step you take concerning *Cyclestheria* is OK, I think. It is understandable that you choose something between the two alternatives. If we one day decide to take the full step of the possible sister group relation to the Cladocera, then a name is already available by Ax (1999). He suggests the term 'Cladoceromorpha.' There are also a couple of new molecular papers out on the issue that seem to support *Cyclestheria* in the mentioned sister group position.

*Submitted by Jørgen Olesen,*
*University of Copenhagen, Denmark*

## BRANCHIOPODA

The quotation from Fryer really encapsulates what is wrong with the old ideas about crustacean phylogeny and taxonomy. This focus on "... animals that work ..." is directly lifted from the later writings of Sidnie Manton. Schram (1993, The British School: Calman, Canon, and Manton and their effect on carcinology in the English speaking world; *Crustacean Issues* 8:321–348) outlined the roots of Mantonian reasoning in an idealist philosophical tradition that passed on through Thompson and his treatise *On Growth and Form*. This is essentially a Platonic view of comparative biology, and stands essentially at odds with the current emphasis, either *a priori* or *a posteriori*, on elucidating ground plans. You are of course free to quote Fryer, but you ought to give fair play to alternative philosophical and conceptual foundations for systematics.

*Submitted by Frederick R. Schram,*
*Zoölogisches Museum, Amsterdam*

## BRANCHIOPODA: ANOSTRACA

Weekers et al. (in press) examined small subunit ribosomal DNA of anostracans from 23 genera belonging to eight of the nine families recognized by Brtek (1997). Their results do not support the family Linderiellidae or Polyartemiidae. Instead, they group *Linderiella* with *Polyartemia* and *Polyartemiella* as a subfamily of the family Chirocephalidae. Morphological considerations support this arrangement in that the three genera share rigid antennal appendages on otherwise simple antennae and double pre-epipodites. Unfortunately, these workers were not able to obtain usable *Artemiopsis*. Thus, the validity of Artemiopsidae remains untested by molecular methods; however, I continue to consider that the morphology of the penes places *Artemiopsis* in the family Chirocephalidae.

### Additional References


Weekers, P. H. H., G. Murugan, J. R. Vanfleteren, and H. J. Dumont. In press. Phylogenetic analysis of anostracans (Branchiopoda: Anostraca) inferred from SSU rDNA sequences. *Molecular Phylogenetics and Evolution.*


*Submitted by Denton Belk,*
*Our Lady of the Lake University,*
*San Antonio, Texas*

## REMIPEDIA

See comments from G. Boxshall under Maxillopoda and from M. Christoffersen under Crustacea.

## REMIPEDIA

In the section about the Remipedia, you mention that the similarities between the Maxillopoda and the Remipedia are symplesiomorphies. But what are these? The only similarities I can think of, I would not consider as symplesiomorphies, but perhaps as convergences. Perhaps it is unwise to mention something like this without also mentioning the characters. The first question people will raise is what these characters are. In the same section you use the term 'basal' about branchiopods, but what does that actually mean? There are two possibilities, either early off split (e.g., sister group) or primitive (or at least with many primitive features), but these are two different things, as addressed earlier.

*Submitted by Jørgen Olesen,*
*University of Copenhagen, Denmark*

## CEPHALOCARIDA

In the section about the Cephalocarida, you say that the sequence of the classes reflects something (it doesn't matter exactly what in this context). My problem here is that I don't think that the sequence of taxa of equal rank in a classification reflects anything. If a classification shall reflect anything concerning relationship, it has to be put into the hierachical categories (like you have done for the classification within the Branchiopoda, for example). I think this is an old way of thinking with no meaning today.

*Submitted by Jørgen Olesen,*
*University of Copenhagen, Denmark*

## MAXILLOPODA

The status of the Maxillopoda remains uncertain. I consider that there is a group of related taxa which form the core of a Maxillopoda: these are the Copepoda, Thecostraca, Tantulocarida and Ostracoda (excluding the Phosphatocopines which are not ostracods and do not even belong to the crown group of the Crustacea). The Mystacocarida and Branchiura may also belong to this group but the available supporting evidence is weaker. I also consider that the Remipedia is related to the maxillopodan lineage. Remipedes share several derived features of the thoracopods, maxillules and maxillae with other maxillopodans as indicated in my paper on comparative musculature (Boxshall, 1997).

### Additional References


Boxshall, G. A. 1997. Comparative limb morphology in major arthropod groups: the coxa-basis joint in postmandibular limbs. In *Arthropod relationships*, eds. R. A. Fortey and R. H. Thomas, 155–167. London: Chapman and Hall.


*Submitted by Geoff Boxshall,*
*Natural History Museum, London*

## MAXILLOPODA

I really understand your difficulties here. To cut the message short, I think you should have chosen to include the component taxa of the Maxillopoda as classes and then skip the 'Maxillopoda' (as you also almost decided to, I can see from your writing).

I know you [are trying] to be conservative by following Bowman and Abele here, but actually, to be real conservative you should skip that level. This would be a choice of the future for the reasons mentioned below.

I think it is better to have your higher level classification to include only what is quite certain. The highest categories (classes) should then be something like the following: Malacostraca, Branchiopoda, Remipedia, Copepoda, Mystacocarida, Branchiura, Thecostraca, Cephalocarida, Ostracoda, Tantulocarida, (Pentastomida).

These are with the highest certainty all monophyletic (not considering that insects may go in somewhere). As for the grouping of these taxa, we appear to know too little yet. As you know, this is reflected in the high number of different schemes put forward that all differ from each other. Perhaps it will take 50–100 years before we get the full story, if ever. The great advantage of having such a flat structure is that it would tell people what the crustacean community thinks is certain, but it would also point at what is unknown by not having any of these weakly supported higher level taxa included (like Maxillopoda, Entomostraca, Thoracopoda, and the one you now suggest being comprised of all non-branchiopod Crustacea). This will be a logical starting point for any students of the Crustacea that want to address the higher level phylogeny. If a taxon like Maxillopoda is included, for example, then the starting point is most likely already polluted.

*Submitted by Jørgen Olesen,*
*University of Copenhagen, Denmark*

## MAXILLOPODA: RHIZOCEPHALA

Boschma (1928) is without any doubt the author of the family Lernaeodiscidae, but both the families Peltogastridae and Sacculinidae must be ascribed to Lilljeborg (1860). This has been duly checked. Boschma lived 1893–1976, and cannot possibly be the author of these two families. Holthuis and I consulted Lilljeborg's (1860) publication, a copy of which is in our library; there is not a shadow of a doubt concerning his authorship!

*Submitted by W. Vervoort,*
*Rijksmuseum van Natuurlijke Historie,*
*Leiden, The Netherlands*

## MAXILLOPODA: COPEPODA

I suggest you strictly adhere to what is already published. Names should in my view not be introduced unofficially but through full and reviewed papers. Two PhD theses have just been completed here with phylogenetic revisions of the Cyclopoida and one branch of Harpacticoida. I could tell you all the changes they entail but that would alter your list quite visibly. The Poecilostomatoida, e.g., are not a separate order but a specialised branch within Cyclopoida. There are many new families and others had to be synonymized. So, please, stick to published and avoid cryptic information (= pers. comm.).

*Submitted by H. Kurt Schminke,*
*Universität Oldenburg, Germany*

## MAXILLOPODA: PENTASTOMIDA

First, on a separate subclass Pentastomida—what can I say. You cite all the relevant papers that argue and provide evidence that these are Branchiura, and yet you reject these and separate them. This is one of the few places where we have good apomorphies to unite the groups involved. If you accept Thecostraca, then why not accept a single subclass Branchiura with two orders: Arguloida and Cephalobaenida?

Concerning the Walossek arguments in the second paragraph: All this Cambrian apparent pentastomid says is that Pentastomida are older than we thought they were. It does not argue against anything. You rightly point out that the fossils might not even be pentastomids. As to whether or not the hosts "were on the scene," you must be careful. Recent issues of *Science* and *Nature* have featured a stunningly preserved early chordate that to all intents and purposes looks like it was drawn by old Al Romer himself when figuring a vertebrate ancestor. This Chengjiang fossil in fact trumps Brusca's suggestion, which is true by the way, that the conodont animal is a chordate.

*Submitted by Frederick R. Schram,*
*Zoölogisches Museum, Amsterdam*

## OSTRACODA

I am sure the classification and appended rationale will be useful and will advance the study of crustaceans. I am still of the opinion that the suborders of the order Podocopida are unnecessary and should be deleted, especially as each contains only one superfamily except for the Cypridoidea, all superfamilies of which are monotypic.

It is likely that the paleontologists will follow the classification that is published in the revised Treatise, and that classification will be determined by Professor Whatley and his team of specialists, which includes Dr. Martens.

As for -acea v. -oidea, you must of course be consistent throughout your classification. Some volumes of the Treatise (most notably the revision of the brachiopods) have now begun to follow the recommendation of the ICZN, but you should realize that these are only recommendations, not rules; and they may sometimes lead to the curious duplications of names among superfamilies and genera. Good luck with the classification. I look forward to seeing the final version.

*Submitted by Roger L. Kaesler, Paleontological Institute, The University of Kansas*

## OSTRACODA

Spelling of Suborder Halocyprina Dana, 1853. Dana (1853: 1281) based his subfamily Halocyprinae and family Halocypridae on his new genus *Halocypris*. Therefore, at least according to present rules, the subfamily should be Halocypridinae and the family Halocyprididae. Dana did not use the names Halocyprina or Halocyprida. If you are basing your Halocyprina and Halocyprida on the family name Halocyprididae, it seems to me that, to be consistent, the suborder should be Halocypridina and the order should be Halocypridida. If you are basing your Halocyprina on the commonly used name for the order, Halocyprida, then I think you are correct in using Halocyprina. Possibly, you should explain your reasoning for using Halocyprina, because I think that you are creating a new spelling for the suborder. [*Editors'note: we retained the spelling Halocyprida for the order, as listed in Bowman and Abele (1982: 13), and Halocyprina for the suborder based on the order name.*]

*Submitted by Louis Kornicker, Smithsonian Institution, National Museum of Natural History*

## STOMATOPODA

I am leery of following suggestions made in abstracts concerning higher taxonomy. Cappola has never published her Pseudosquilloidea (which I see you accept) with documented reasons for her decision. In fact, some of the new analyses of Ahyong and Hof (not yet published) would not entirely support such an arrangement.

Thus, while we are at it, you need to turn to [page 86 in original draft]. I suggest for now you simply leave all the "gonodactyloid" families in one superfamily Gonodactyloidea. When we can identify clear clades and suggest valid groupings, you can change it; or when people actually publish revisions in a refereed journal.

*Submitted by Frederick R. Schram, Zoölogisches Museum, Amsterdam*

## AMPHIPODA

Although I agree in general with the thrust of your arguments, you fail to recognise the complexity of amphipod morphology and the lack of family level revisions, which makes the development of an acceptable classification extremely difficult. Suborder and families were established long ago and for the most part have never been revised. Superfamilies were to a certain extent based on gestalt, which worked well for some groups like corophioids, lysianssoids and haustorioids, but failed for families which didn't show clear body-plan relationships. Even groups as seemingly distinctive as the Lysianassoidea are very difficult to define morphologically when all genera are considered. When Barnard and Karaman (1991) collapsed the majority of corophioid families, they did it because these traditional families (although workable when they were originally established) were no longer definable and could no longer be supported. Genera described over the years had been pigeon-holed into one family or another until any characters which might define them had become totally diluted. It will take a large effort using modern phylogenetic techniques to develop an acceptable classification. The results of these works have to be published in reputable journals after careful peer-group review. Attempts to revise classifications are underway. For instance, Lowry and Myers are currently revising the iphimedioid group and Myers and Lowry are revising the corophioid group. The website www.crustacea.net has recently been established to publish information and retrieval systems (electronic monographs) for all crustaceans. For instance, Watling and his students are currently preparing cumacean data bases and Lowry and his students are working on amphipod data bases for the website. It is unfortunate that the use of poorly refereed journals and pseudophylogenetic methodologies have been used in some cases to produce untestable and, in some cases, unacceptable classification systems.

Because of these problems, we currently list our taxa alphabetically in the Amphipoda. I do not see the problem. All classifications are hypotheses which change as new hypotheses are produced. In a large monograph, it is fine to discuss and list the phylogenetic classification, but probably the taxonomic section should be alphabetical. Trying to find families or genera listed phylogenetically in a large monograph can be a nightmare for those not in the know (basically everyone but experts). It is relatively easy, for example, to find a family level taxon in Barnard and Karaman (1991). One does not have to continually consult the index.

*Submitted by Jim Lowry, Australian Museum, Sydney*

## AMPHIPODA: GAMMARIDEA

As Ed Bousfield was not present at the amphipod conference in Amsterdam to defend the value of phyletic vs. alphabetical classification of the Gammaridea, several points raised in the Vader-Baldinger-K-S-Watling report seem largely matters of mechanics rather than matters of phyletic substance.

Some points of your recent "critique" summary may require modification, viz: (1) "the schedules of Jerry Barnard and Ed Bousfield (are) often not very compatible" and (2) ". . . not espousing one worker's view over another." With all due respect to Jerry's enormous contribution to gammaridean taxonomy, his formal "track record" in gammaridean phylogeny was actually quite modest in scope.

Thus, he did recognize (temporarily, at various times) Talitroidea Bulycheva, 1957, Corophioidea Barnard, 1973, and Haustorioidea Barnard and Drummond, 1982. Several of Jerry's informal "anglicized" groupings of freshwater families (e.g., "gammarida," "crangonyctoids," "hadzioid group," etc., in Barnard and Barnard, 1983; Williams and Barnard, 1988) rather closely resemble some of the superfamilies (and families) formally named and fully defined previously (1973, 1977, 1979, 1982) by Bousfield and co-workers (e.g., about 75% compatibility with Gammaroidea, Hadzioidea, Crangonyctoidea, Melphidippoidea, etc.). However, he did not attempt formal phyletic groupings of most marine gammaridean families, nor formal integration with other amphipod suborders. Unlike Sars (1895), Stebbing (1906), and other "turn-of-the-century" workers, Jerry apparently did not recognize the significance of reproductive form and behaviour in amphipod phylogeny. Jerry's final major work (with Gordan Karaman, 1991, p. 7) disavowed the significance or use of the formal superfamily concept, and listed families alphabetically rather than phyletically or semi-phyletically (as in Sars and Stebbing, above). Some classifications are based on carefully defined characters and character states that have required (and will continue to require) modification according to features found in subsequently discovered species and genera, and are consistent at proper classificatory levels. The cladistic arrangement by Kim and Kim (1993), reviewed rather unfavourably by Schram (1994), underscores the unreliability of cladistic analysis when care is not taken in the appropriate selection and accurate definition of characters and character states.

[*Concerning your statement about Bousfield and Shih*], Bousfield and Shih (1994) represents an updating and refinement of previous ~20 years of study and publication on gammaridean phylogeny.

[*Concerning your statement about Reptantia*], "Natantia" and "Reptantia" are terms (names) pragmatically defined, but not incorporated formally by Bousfield and Shih (1994). The terms are analogous to former groupings of families and superfamilies, etc., within the Order Decapoda.

[*Concerning your statement about names and dates*], omission of author names and dates in tabular listing of families and superfamilies is modeled after similar "heading" omissions in Barnard's "Families and Genera . . ." (1969) and earlier "Index . . ." (1958). Obviously, these names are fully treated in the major references (e.g., Stebbing, 1906; Gurjanova, 1951; Bousfield 1979, 1982, 1983; Schram, 1986). Readers are expected to provide something of substance to the discussion, such as commentary on the paper's extensive analysis of "across-the-phyletic-board" variability of major characters and character states (antennae to telson) that would be of prime significance in a cladistic treatment.

[*Concerning your statement to the effect that we presented different phylogenetic hypotheses in our 1994 paper*], Bousfield and Shih acknowledge (problems in resolution) that they do not have a "final answer" to the probably correct evolutionary history of the Amphipoda (only one answer can be correct!). Their "semi-phyletic" methodology modifies the strictly phenetic format of Sneath and Sokal (1973) by careful ordering of character states to arrive at a "plesio-apo-morphic index" of probably correct phyletic relativity for each taxon. This approach tends to minimize the negative effects of homoplasious convergence in many of these character states (analyzed above). Mike Ghiselin (1984) correctly points out, rigid and uncritical application of cladistic methodology alone quite frequently leads the user to a less-than-credible phylogenetic result. Thus, use of the "Wagner 78" cladistic program often provides multiple "trees" from the same data base, each one different, each one tending to invalidate the other, and none probably correct!

[*Concerning your statement about cladistic analyses having high priority*], to my knowledge, cladistic "purists" have not yet actually demonstrated a cladistically derived treatment of all 118 gammaridean families of your list. Chances of doing so would appear "slim-to-non-existent." Instead, advocacy of rDNA methodology would probably result much sooner in a most-probably-correct answer!

[*Concerning your statement that most workers would prefer to see the families listed alphabetically rather than by superfamily*], how surprising that such an unsupported statement should come from Les Watling, a confirmed crustacean phylogenist! On more serious reflection, Les may find that quite a few current workers (e.g., Mike Thurston, John Holsinger) do not "give up" so easily on the full solution of this difficult problem.

[*Concerning your use of the word hypotheses*], do you mean "concepts"? All family and superfamily names represent "concepts" of presumed natural groupings of species. Some are better defined (in terms of careful definition of character states) and longer time-tested than others. Most superfamily names in Bousfield and Shih (1994) have been carefully and fully (multiple-character) defined, their component families named, and time-tested (by other workers as well) over a 15+ year period. Since superfamily taxonomic stability (75%) would appear at least equal to that of the component family-level names of the current Martin–Davis list, although both lists are "conceptual," neither can realistically be termed "hypothetical."

## Additional References

[*Note: Dr. Bousfield did not supply references to all papers mentioned above.*]

Bousfield, E. L. 1995. A contribution to the natural classification of Lower and Middle Cambrian arthropods: food gathering and feeding mechanisms. *Amphipacifica* II(1):3–34.

Bousfield, E. L. 1996. A contribution to the reclassification of neotropical freshwater hyalellid amphipods (Crustacea: Gammaridea: Talitroidea). *Bull. Mus. civ. St. nat. Verona* 20[1993 (1996)]:175–224.

*Submitted by Ed Bousfield, Ottawa, Canada*

### AMPHIPODA: GAMMARIDEA

There would seem to be a second main reason why you might regret not employing a natural (superfamily) classification of the Gammaridea. Not only the Lysianassoidea, Talitroidea and Corophioidea, but about 75% of superfamilies of the Bousfield–Schram phyletic classification (including Jerry Barnard's anglicized versions) are variously utilized by major workers today—if only because they make pragmatic (workable) sense.

Interestingly, and to my knowledge, none of those who apparently condemn the present superfamily categories because they "have not been derived cladistically" has attempted a natural treatment of all 113 families (embracing ~5000+ species!) of your list, based on cladistics alone. Why?—not only is the task extremely difficult and time-consuming, but the feasibility of obtaining a single, credible, "all-inclusive" answer with that methodology alone is highly improbable, and I think they know it! On the other hand, rDNA studies seem virtually unaffected by homoplasious convergence of morphological character states "across the board" and are quite promising—if only someone would get started!

The second, and perhaps more important, essentially scientific reason is that gammarideans, virtually alone among crustacean higher taxa (including the 3 other amphipod suborders!) would remain unclassified phyletically. Such an anomalous situation will be corrected inevitably—hopefully sooner than later—providing the principal reason for phyletic classification in the forthcoming CNAI lists and Pacific amphipod guide. Sars, Stebbing, and other perceptive "turn-of-the-century" amphipodologists might then cease "rolling over in their graves"!

*Submitted by Ed Bousfield, Ottawa, Canada*

### ISOPODA

In the near future, we must abandon the use of Linnean categories, because we are currently identifying many more encaptic levels of monophyletic groups than there are hierarchical levels in the Linnean system. The Paranthuridae, for example, are definitely a monophyletic group that contains further subgroups. To erect new families for these subgroups means to give up a categorical rank for the taxon Paranthuridae.

The same problem exists for the Epicaridea. New molecular evidence (Ph.D. thesis of H. Dreyer) proves that these parasites of crustaceans are derived from a common ancestor shared with the Cymothoidae (fish parasites). Thus, the suborder Epicaridea is placed within the suborder "Flabellifera" or, more precisely, within the suborder Cymothoidea sensu Wägele (1989), the sister group of the suborder being a taxon classified as a family.

Concerning the hypothesis that the Sphaeromatidae, Serolidae, and other groups are derived from a disc-shaped ancestor (the ancestor of the Sphaeromatidea sensu Wägele, 1989), new evidence was discovered with the fossil *Schweglerella stroebli* (Polz, H. 1998. *Archaeopteryx* 16:19–28). This animal shows neither the apomorphies of the Serolidae nor of the Sphaeromatidae or other related extant taxa, but shows those characters identified as apomorphies of the suborder Sphaeromatidea (e.g., disc-shaped body, head immersed in first pereonite, dorsal eyes).

The subdivision of the Oniscidea into Tylomorpha and Ligiamorpha does not reflect the phylogeny of terrestrial isopods, as shown by Erhard (1996, 1998). Detailed phylogenetic analyses based on morphological characters will be published soon (Ph.D. theses of C. Schmidt and of A. Leistikow).

*Submitted by J. W. Wägele, Ruhr-Universität Bochum, Germany*

### SYNCARIDA

The author of both the Bathynellidae and Bathynellacea is Chappuis, 1915. I have copied the paper by Chappuis (1915) for you. I am a bit surprised that you cite Lopretto and Morrone (1998) who have added nothing new to our understanding of Syncarida. You should quote those who have.

*Submitted by H. Kurt Schminke, Universität Oldenburg, Germany*

### DECAPODA: CARIDEA

I am puzzled to find the family Barbouridae among the superfamily Bresilioidea. Chace (1997) put them among the hippolytids. Christoffersen (1987, 1990) put them in the superfamily Crangonoidea. Who put them among the bresilioideans, and why? This is not stated clearly in your section on the superfamily Bresilioidea on p. 61. [*Editor's note: the family Barbouriidae Christoffersen was mistakenly placed by us in the Bresilioidea; this has since been corrected and they are now listed among the Alpheoidea.*]

Otherwise, the classification contains the usual fights between lumpers and splitters. I think that Christoffersen's classification may fall apart in the future because much of it is based on descriptions from the literature and not on examination of actual specimens. Some of the descriptions are inaccurate or do not contain pertinent information needed in classification today.

*Submitted by Mary K. Wicksten, Texas A&M University*

## DECAPODA: CARIDEA

I of course must strongly disagree with the proposed arrangement of the caridean families into superfamilies, because I see this as a retrocess from taxa sustained by apomorphic characters (Christoffersen, 1990) back to groupings based on overall resemblance, authority (Chace, 1992; Holthuis, 1993), or arbitrary usage. It is true that my proposals have had little following in the carcinological community, and that some of my employed characters may be questionable. But it is also true that my efforts remain the first attempt to produce a phylogenetic system of the Caridea. Because my system differs substantially from the traditional arrangements, my suggestions have usually been dismissed as totally heretical, without any serious attempt to argue alternative possibilities sustained by better uniquely shared characters. It is rather depressing to note that the present authors follow this same tactic. They do not accept a single superfamily as synthesized in Christoffersen (1990). More explicitly, but without justification, they reject my proposal to combine alpheoids, crangonoids and pandaloids into a monophyletic taxon. This is surprising to me, because these superfamilies, as redefined in my cited works, share a remarkable synapomorphy, the multiarticulated carpus of the second pereiopod, which is a unique adaptation within the carideans for body cleaning. For this transformation series, there is even a transitional stage represented by the nematocarcinoids, in which the carpus of the second pereiopods is longer than in the preceding carideans, before being subdivided in the sister group represented by pandaloids, crangonoids, and alpheoids. At a still higher level of generality, this transformation series is congruent with the presence of a well developed incisor process on the mandible of palaemonoids and all the previously mentioned superfamilies. Going to a lower hierarchical level, there is further congruence with the uniquely expanded first cheliped in crangonoids and alpheoids. My rearrangements of the traditional families into superfamilies eliminate all the paraphyletic family-level taxa, including the notably unsatisfactory Hippolytidae. Finally, just to mention one remarkable autapomorphy justifying one of my new superfamilies, only palaemonids and rhynchocinetids share a second distolateral tooth on the basal segment of the antennule, in addition to the usual stylocerite. Some researchers complain that I presented few characters for each node, but this is because my approach is qualitative and I selected the best possible evidence from detailed studies of the total morphological and species diversity of the Caridea. To refute the phylogenetic system, it is necessary that researchers argue for alternative replacement characters where they believe I have failed. Simply ignoring the system does not justify the usual assumption that my arrangements are totally wrong!

*Submitted by Martin L. Christoffersen,*
*Federal University of Paraíba, Brazil*

## DECAPODA: REPTANTIA

I really do not understand why you do not use a separate category for Reptantia. It is one of the clearest, most universally accepted groups (taxonomically or cladistically) among the decapods that we have.

*Submitted by Frederick R. Schram,*
*Zoölogisches Museum, Amsterdam*

## DECAPODA: ASTACIDEA

You really do get yourselves into deep water when you try to offer editorial comments on cladistic analyses. Here you hit another one. Where do you get the idea of "extremely primitive *Neoglyphea*" from? Forest and de St. Laurent (1989, Nouvelle contribution a la connaissance de *Neoglyphea inopinata* a propos de la description de la femelle adulte, *Res. Camp. Musorstom 5,* Memoirs Mus. Nat. His. Nat., series A, 144:75–92) made [a] good argument for allying glypheoids with astacids—not a particularly *primitive* alliance. My own preliminary examination of decapod phylogeny (submitted, *Hydrobiologia*) not only fairly well confirms the Scholtz and Richter scheme, but also squarely places *Neoglyphea* within the Fractosternalia.

As I say, my own examination of the subject in connection with an assignment to address decapod phylogeny in connection with the beginning revision of the decapod section of the *Treatise on Invertebrate Paleontology* has, to my surprise, uncovered the basic robustness of the Scholtz and Richter analysis. I think you would do well to leave yourself an opening here.

Of course, I see why you are keen to downplay Scholtz and Richter because here you adapt a very conservative combination of "clawed lobsters." I can accept this for now. However, I think it would only be fair for you to point out that Scholtz and Richter would segregate the "clawed (true) lobsters" as Homarida from the crayfish as Astacida. My own on-going, recent work indicates that at least the genus *Neoglyphea* is a fractosternalian in some kind of proximity to the Astacida, and that *Enoplometopus* may even be a separate clade from the Nephropoidea. That this paraphyly should emerge among "lobsters" is not too surprising, since we discover again and again that supposedly robust, traditional groups bearing a lot [of] plesiomorphies emerge on closer examination as paraphyletic taxa. Why should macrurous lobsters be any different?

*Submitted by Frederick R. Schram,*
*Zoölogisches Museum, Amsterdam*

## DECAPODA: ANOMURA

I fully agree with the different parts of my specialty (Anomura). I agree with the changes included in this new version. As you mention . . . we need more studies (especially molecular) to improve our knowledge on the phylogeny and the classification of Crustacea, and obviously new, and perhaps strong, changes will come in the near future. However, we need to put in order our present knowledge of the group.

*Submitted by Enrique Macpherson, Centre D'Estudios Avancats de Blanes, Spain*

## DECAPODA: ANOMURA

I can only address your classification of the Anomura. Forest (1987a, b), while concurring with McLaughlin's (1983) argument that the Paguridea represented a monophyletic taxon, did not agree with her elimination of the Coenobitoidea as a superfamily. Consequently he elevated the Paguridea to rank of Section and reinstated the superfamily Coenobitoidea to include the families Pylochelidae, Diogenidae and Coenobitidae. He did concur with McLaughlin's removal of the Lomidae and its elevation to superfamily. He did not address the hierarchical ranking of the other Anomuran superfamilies. McLaughlin and Lemaitre (1997) acknowledged Forest's sectional ranking for the Paguridea, but continued to refer to all of the anomuran major taxa as superfamilies. However, Forest et al. (2000), Forest and McLaughlin (2000), and de Saint Laurent and McLaughlin (2000) all refer to the superfamilies Coenobitoidea and Paguroidea, under the Section Paguridea.

I personally still believe that the Paguridea represent a monophyletic taxon; however, I also believe that Forest's argument for reinstatement of the Coenobitoidea is valid. For hierarchical balance within the Anomura, perhaps the other superfamilies should similarly be elevated to Section rank in your classification.

### Additional References


Forest, J. 1987a. Les Pylochelidae ou "Pagures symetriques" (Crustacea Coeno-bitoidea). In *Résultats des campagnes MUSORSTOM. Mémoires du Muséum National d'Histoire Naturelle,* série A, *Zoologie,* vol. 137, 1–254, figs. 1–82, plates 1–9.

———. 1987b. Ethology and distribution of Pylochelidae (Crustacea Decapoda Coenobitoidea). *Bulletin of Marine Science* 41(2):309–321.

Forest, J., M. de Saint Laurent, P. A. McLaughlin, and R. Lemaitre. 2000. The marine fauna of New Zealand: Paguridea (Decapoda: Anomura) exclusive of the Lithodidae. *NIWA Biodiversity Memoir* 114 (in press).

Forest, J., and P. A. McLaughlin, 2000. Superfamily Coenobitoidea. In *The marine fauna of New Zealand: Paguridea (Decapoda: Anomura) exclusive of the Lithodidae,* eds. J. Forest, M. de Saint Laurent, P. A. McLaughlin, and R. Lemaitre. *NIWA Biodiversity Memoir* 114.

McLaughlin, P. A. and R. Lemaitre. 1997. Carcinization in the Anomura—fact or fiction? I. Evidence from adult morphology. *Contributions to Zoology, Amsterdam* 67(2):79–123, figs. 1–13.

Saint Laurent, M. de, and P. A. McLaughlin, 2000. Superfamily Paguroidea, Family Paguridae. In *The marine fauna of New Zealand: Paguridea (Decapoda: Anomura) exclusive of the Lithodidae,* eds. J. Forest, M. de Saint Laurent, P. A. McLaughlin, and R. Lemaitre. *NIWA Biodiversity Memoir* 114.


*Submitted by Patsy McLaughlin, Shannon Point Marine Center, Anacortes, Washington*

## DECAPODA: BRACHYURA

As before, I think that the Oregoninae of Garth should be elevated to a family. I contacted Michel Hendrickx about the classification. He in turn quoted a paper that provided larval evidence for the distinction of the group as a family, and said that he will treat the group as such in his forthcoming work on crabs. Please contact Michel for further information. If you cannot contact him, let me know and I'll find that larval paper for you. My own suspicion is that the oregoniids are not covered in most monographs because they are a cirumArctic and boreal northern hemisphere group that does not range at all into tropical waters, where most researchers work!

*Submitted by Mary K. Wicksten, Texas A&M University*

## DECAPODA: BRACHYURA

I strongly believe that the Pinnotheridae are not monophyletic. So if I argued that this family "should remain in the Thoracotremata based on evidence from DNA sequencing" [as cited in your classification], I should add that this might only be true for some of its constituent subfamilies or genera. My statement was made based on the phylogenetic position of *Pinnixa* in molecular analyses that showed a strikingly close relationship to the Ocypodinae (Schubart et al., 2000a).

I also think that the Ocypodidae in the traditional sense as well as the Ocypodoidea as defined in the latest draft of your classification might not be monophyletic. Molecular as well as larval morphological data suggest a close relationship between the Varunidae (Grapsoidea) and the Macropthalminae (Schubart et al., 2000a; Schubart and Cuesta, unpublished). I think that this possible phylogenetic link would be another reason to elevate ocypodid subfamilies to family level as already considered in your draft and suggested for the Grapsidae (Schubart et al., 2000b). This would certainly make justice to ocypodoid morphological diversity and allow a more objective comparison with other thoracotremes in the future.

I disagree on the use of the superfamily name "Grapsidoidea." Since the stem of the name is Graps- (based on *Cancer grapsus* Linnaeus, see also family name Grapsidae) and the ending for superfamilies is -oidea, the superfamily should be called Grapsoidea (and not Grapsidoidea). The fact that the term Grapsoidea has been used in the past for a much wider systematic grouping of eubrachyuran crabs and is now restricted to the families Grapsidae, Gecarcinidae, Plagusiidae, Searmidae, and Varunidae should not influence the nomenclature.

### Additional References


Schubart, C. D., J. A. Cuesta, R. Diesel, and D. L. Felder. 2000b. Molecular phylogeny, taxonomy, and evolution of non-marine lineages within the American Grapsoidea (Crustacea: Brachyura). *Molecular Phylogenetics and Evolution* (in press).

Schubart, C. D., J. E. Neigel, and D. L. Felder. 2000a. The use of the mitochondrial 16S rRNA gene for phylogenetic and population studies of Crustacea. *Crustacean Issues* 12 (in press).



*Submitted by Christoph Schubart,*
*Universität Regensburg, Germany*


### DECAPODA: BRACHYURA

Although recently I published my arrangement of the brachyuran families, I have some new discoveries in the brachyuran classification, but it is not finished and it will be published next year. I was able to classify all dromiacean families into superfamilies, but not the eubrachyuran ones, because there are many families with obscure systematic position: Orithyiidae, Calappidae, Matutidae, Astenognathidae, Hexapodidae, Palicidae, Dairodidae and many up to now undescribed families (Acidopidae, Melybiidae, Speocarcinidae, etc.). Here are some of my remarks.

(1) Dynomenidae are the most primitive Dromioidea, because only the last pair of legs is aberrant. (2) Among Homoloidea, the Poupinidae are the most primitive because the last pair of legs are of "normal" structure but are partly subdorsal in position. (3) Raninidae are "Podotremata" (i.e. Dromiacea) because their sexual openings in both sexes are on the coxae of the legs (hence the name Podotremata). (4) The most primitive eubrachyuran family is the Atelecyclidae, because they have the antennules and antennae longitudinally directed, a narrow thoracic sternum, thoracic sternites 4/5–7/8 continuous (entire), and sternites nearly regularly metamerized. (5) The Dorippidae are highly derived and aberrant: the dorsal position of the posterior pair of legs, the sternite 8 facing dorsally, and the narrowed buccal cavern all are secondarily attained. The similarity with the Dromiacea is thus superficial. (6) The same could be said for the Leucosiidae: highly derived crabs and consequently should be placed at the end of the classificatory scheme of the Heterotremata. (7) The Majidae are only one family with many subfamilies. The arrangement is enclosed [Števčić, Z. 1994. Contribution to the re-classification of the family Majidae. *Periodicum Biologorum* 96:419–420]. (8) The Parthenopidae are more primitive than Majidae, and therefore should be ahead of the Majidae. (9) The Retroplumidae are a very derived brachyuran family. (10) Geryonidae have a similar organization to the Goneplacidae s.s. (11) Your Xanthoidea is a highly polyphyletic group. (a) The most primitive "xanthoids" are the Eriphiidae, not Menippidae! The most primitive Eriphiidae have sternites 4/5–7/8 entire, abdominal segments freely articulated in both sexes, and the second gonopod longer than the first. They are probably related to Trapeziidae. In the same assemblage with the Eriphiidae are the Pilumnoididae Guinot and Macpherson, 1987. (b) Xanthidae s.s. have [some] primitive representatives (Krausinae, with sternal sutures 4/5–7/8 entire), but abdominal segments 3–5 in the male are fused, and the second gonopod is short. They are related to the Panopeidae/Panopeinae and the Pseudorhombilidae. (c) Pilumnidae have a primitive abdomen (all segments freely articulated in both sexes) but specific first and second gonopods, the latter short. They are related to the Eumedonidae (in fact the Eumedoninae). (d) Goneplacidae s.s. are in fact a very small taxon, without any close relationships with the Xanthidae. They are probably close to the Geryonidae and Euryplacidae/Euryplacinae. (12) The Potamidae are in fact a very difficult problem, however the gaps among subfamilies are not quite distinct. The gaps are not always [clear] and therefore the separation of the freshwater crabs into families remains uncertain. (13) I think that between Ocypodidae and Mictyridae and between Grapsidae and Gecarcinidae the gaps are not decisive and only Ocypodidae and Grapsidae are true families (this will be published later). (14) Finally, I think that the Cancroidea are not a taxon, they are only a grade, not a clade (taxon i.e., monophyletic group). (15) Hepatinae are a subfamily of the family Aethridae. (16) Palicidae belong to the Heterotremata, with no close affinity with the Ocypodidae.


*Submitted by Zdravko Števčić,*
*Rudjer Boskovic Institute, Croatia*


### DECAPODA: BRACHYURA

Concerning my special knowledge, the Brachyura, I do not agree with all decisions (see my responses), but I respect them. May I add my feeling, however. Concerning the Podotremata, the molecular data seem to outweigh all other considerations, despite the fact that the first results (Spears and Abele, 1988; Spears, Abele, and Kim, 1992) were fragmentary, based only on very few taxa (only two Dromiidae were studied; and the conclusion was made without any Dynomenidae, Homolodromiidae, Homolidae, Latreilliidae, Cyclodorippidae, Cymonomidae, nor Phyllotymolinidae) and that the new results are not yet published. I am happy to see that Spears now returns to the opinion that the Dromiidae are true Brachyura, but we wait her paper where the new demonstration is given.

Concerning your Section Raninoida, you write (p. 66, 69) that there is "possibly a mistake." I recognize that the problem of the placement of on the one hand Cyclodorippidae, Cymonomidae, and Phyllotymolinidae, and on the other hand the Raninoidea is difficult, because they do not clearly enter in a major group. You write that, for Spears herself (p. 69), "molecular data seem to indicate a placement [of Cyclodorippoidea] somewhere between the raninids and the higher eubrachyurans." So, the molecular data exactly give the same results that the morphological and ontogenetic ones. The two groups Raninoidea and Cyclodorippoidea (the last name is used by convenience, but perhaps they form three distinct families, see Tavares) seem apart, but where is the best way?

*Submitted by Danièle Guinot, Muséum National d'Histoire Naturelle, Paris*

## DECAPODA: BRACHYURA

Evidence from morphology and larval development points to the polyphyletic nature of the Trapeziidae. There are three separate groups: one comprises *Trapezia, Quadrella, Hexagonalia, Calocarcinus, Philippicarcinus* and *Sphenomerides*, a second *Tetralia* and *Tetraloides*, and a third *Domecia, Jonesius, Palmyria* and *Maldivia.*

*Submitted by Peter Castro, California State Polytechnic University, Pomona*

## DECAPODA: BRACHYURA

I disagree that all Brachyura with female gonopores on P3 coxa and with spermathecae at the extremities of thoracic sutures 7/8 are separated in two different major sections, Dromiacea and Eubrachyura, with the Raninoidea and Cyclodorippoidea distributed in a basal group inside the Eubrachyura. In that case, how to make a definition of both Dromiacea and Eubrachyura as a whole? The Podotremata may receive all Brachyura with female gonopores on P3 coxa and with spermathecae at the extremities of thoracic sutures 7/8, i.e., two different apertures. The Eubrachyura may receive all Brachyura with a sternal location of female gonopores (vulvae on the thoracic sternum, sternite 6); there is now a sole female orifice for reproduction (egg laying, intromission of male pleopod, and storage of the spermatozoas). Another synapomorphy (among others) of the assemblage Heterotremata-Thoracotremata is the morphology of the first male pleopod, which is completely closed and provided with two distinct basal foramina (instead of only one in the Podotremata). To concile the evident apart position of the Raninoidea and Cyclodorippoidea (but, perhaps consider three distinct families: Cyclodorippidae, Cymonomidae, Phyllotymolinidae), I suggest to range them among the Podotremata in Archaeobrachyura Guinot, 1977 emend. (i.e. with the exclusion of the Homoloidea).

*Submitted by Danièle Guinot, Muséum National d'Histoire Naturelle, Paris*

## DECAPODA: BRACHYURA: DROMIACEA

I disagree that the section Dromiacea contains the Homoloidea. The Dromiacea and Homoloidea are two different lineages. I suggest to consider a Section Podotremata, with three subsections: Subsection Dromiacea, containing two superfamilies Homolodromioidea (Homolodromiidae) and Dromioidea (Dromiidae, Dynomenidae); Subsection Homoloidea (Homolidae, Latreilliidae, Poupiniidae); Subsection Archaeobrachyura (Cyclodorippidae, Cymonomidae, Phyllotymolinidae, and Raninidae). The monophyly of the Dromiacea is well supported by many features; the same for Homoloidea. I recognize that the monophyly of the Archaeobrachyura emend. (without the Homoloidea) is not so well supported and that these crabs show puzzling features, but they are all very specialized and modified by the burrowing life. Their attribution to the Podotremata is, at least for the moment, supported by the appendicular location of female gonopores (on P3 coxa) and the spermathecae at the extremities of thoracic sutures 7/8, the features of the sternal plate, the arthrodial cavities of the pereiopods, and others characters. If we include the Cyclodorippidae, Cymonomidae, Phyllotymolinidae, and the Raninidae in the Eubrachyura, which becomes the diagnosis of the Eubrachyura?

*Submitted by Danièle Guinot, Muséum National d'Histoire Naturelle, Paris*

## DECAPODA: BRACHYURA: HETEROTREMATA, THORACOTREMATA

It is important to recall the original definition of the taxa given by Guinot (1977, 1978).

> The section Herotremata contains the Brachyuran families, ALL THE MEMBERS of which are sternitreme for the female gonopores, and ONLY some members, at least, are podotreme for the male gonopores.
>
> The section Thoracotremata contains the Brachyuran families, all the members of which are sternitreme for the female and male gonopores. It means that, for the Heterotremata, in the Leucosiidae or Leucosioicea by example it exists members with male gonopores on the P5 coxa and other members with sternal male apertures. But, in the last case, it is only a coxo-sternal location of the penis. The same is true for the Dorippidae, where some members show a coxo-sternal location of the penis.

*Submitted by Danièle Guinot, Muséum National d'Histoire Naturelle, Paris*

# APPENDIX II. LIST OF CONTRIBUTORS

The following are colleagues who graciously gave of their time to review various drafts of the Classification of Recent Crustacea.

Abele, Lawrence G.
Baba, Keiji
*Belk, Denton
Bousfield, Ed
Boxshall, Geoff
Brandt, Angelika
Brendonck, Luc
Briggs, Derek
Brusca, Gary
Brusca, Richard
Cadien, Don
Camp, David
Castro, Peter
Causey, Douglas
Chace, Fenner
Christoffersen, Martin
Clark, Paul
Cohen, Anne
Crandall, Keith
Crosnier, Alain
Cumberlidge, Neil
*Dahl, Erik
Dahms, Hans-Uwe
Davie, Peter
Elofsson, Rolfe
Felder, Darryl L.
Feldmann, Rodney
Felgenhauer, Bruce
Fryer, Geoffrey
Galil, Bella
Grygier, Mark
Guinot, Danièle
Haney, Todd
Harvey, Alan
Hayashi, Ken-Ichi
Heard, Richard
Hendrickx, Michel
Hessler, Robert
Ho, Ju-Shey
Høeg, Jens
Hof, Cees
Holsinger, John
Holthuis, Lipke B.
*Humes, Arthur
Huys, Rony
Jamieson, Barry
Jones, Diana S.
Kaesler, Roger
Kensley, Brian
Kornicker, Lou
Larsen, Kim
LeCroy, Sara
Lemaitre, Rafael
Lowry, Jim
MacPherson, Enrique
Maddocks, Rosalie
*Manning, Ray
Markham, John
McLaughlin, Pat
Mickevich, Mary
Modlin, Richard
Morgan, Gary
Myers, Alan
Newman, William
Ng, Peter
Olesen, Jørgen
Poore, Gary
Regier, Jerome C.
Rice, Tony
Richer de Forges, Bertrand
Richter, Stefan
Riley, John
Sakai, Katsushi
St. Laurent, Michele de
Schminke, Horst
Scholtz, Gerhard
Schram, Frederick
Secretan, Sylvie
Schubart, Christoph
Sorbe, Jean Claude
Spears, Trisha
Števčić, Zdravko
Takeuchi, Ichiro
Tavares, Marcos
Thomas, James D.
Tudge, Christopher
Vereshchaka, Alexander
Vervoort, W.
Wägele, Wolfgang
Wallis, Elycia
Walossek, Dieter
Watling, Les
Whatley, Robin
Wicksten, Mary K.
*Williams, Austin
Wilson, Buz
Yager, Jill
Young, Paulo
Zimmerman, Todd L.

* Denotes researchers recently deceased.

# APPENDIX III. OTHER CRUSTACEAN RESOURCES

This appendix is subdivided into four sections. Section III-A contains a list of journals and newsletters and their current editors and addresses. Section III-B is an alphabetical list of currently active web sites and their URLs, followed by a short selection of "personal pages" of some workers with crustacean information on their web sites. Section III-C is a list of crustacean-related listservers. Section III-D is a list of natural history museums with significant crustacean holdings, some of which have searchable crustacean databases.

## III-A. JOURNALS AND NEWSLETTERS

### 1. JOURNALS

Journals that publish only crustacean-specific articles are rather few and currently include only the following (listed alphabetically).

*Crustaceana*

Description: "International Journal of Crustacean Research" publishing "papers dealing with Crustacea, from all branches of Zoology." Issued eight times per year (January, February, March, April, June, July, September, October, November, and December).

Current editor and address: J. C. Von Vaupel Klein, Crustaceana, Editorial Board Administrative Office, Beetslaan 32, NL-3723 DX, Bilthoven, The Netherlands.

Publisher: Brill Academic Publishers, Inc., Leiden, The Netherlands.

*Crustacean Issues*

Description: An irregular series of collections of papers on Crustacea, each published as a hardbound volume and covering a discrete crustacean topic.

Current editor and address: General editor, Frederick R. Schram, Zoological Museum, University of Amsterdam.

Publisher: A. A. Balkema, Rotterdam, The Netherlands.

*Crustacean Research* (formerly *Researches on Crustacea*)

Description: A publication of the Carcinological Society of Japan, publishing papers dealing with "any aspect of the biology of Crustacea." Issued quarterly.

Current editor: Keiji Baba, Crustacea Research, Faculty of Education, Kumamoto University, 860–8555, Japan.

Publisher: Carcinological Society of Japan and Shimoto Printing, Kumamoto.

*Journal of Crustacean Biology*

Description: The official journal of The Crustacean Society, "for the publication of research on any aspect of the biology of Crustacea." Issued quarterly.

Current editor: David K. Camp, Journal of Crustacean Biology, P.O. Box 4430 Seminole, Florida 33775–4430, USA.

Publisher: The Crustacean Society and Allen Press, Lawrence, Kansas.

*Nauplius (Revista da Sociedade Brasiliera de Carcinologia)*

Description: The journal of the Sociedade Brasileira de Carcinologia, publishing "original papers based on research in any aspect of crustacean biology, including taxonomy, phylogeny, morphology, development, physiology, ecology, biogeography, bioenergetics, aquaculture and fisheries biology." Issued quarterly.

Current editor: Mónica A. Montú, Nauplius, Laboratorio de Carcinologia, Departamento de Oceanografia—FURG, Caixa Postal 474, CEP 96201–900, Rio Grande, RS, Brazil.

Publisher: Sociedade Brasileira de Carcinologia.

There are of course many more journals that publish taxonomic/systematic/phylogenetic studies of crustaceans along with papers on other invertebrate groups. We conducted an informal survey of the subscribers to the crustacean listserver CRUST-L in March of 2000 and asked members to name the journals they consult on a regular basis for new information on crustacean relationships. The following journals, arranged alphabetically, were all mentioned more than once in that survey: *Acta Zoologica, Arthropoda Selecta, Biological Bulletin, Bulletin of Marine Science, Canadian Journal of Zoology, Comptes Rendus de l'Academie des Sciences, Contributions to Zoology* (University of Amsterdam), *Deep-Sea Research, Evolution, Fishery Bulletin* (US), *Fossils and Strata, Gulf and Caribbean Research* (formerly *Gulf Research Reports*), *Hydrobiologia, Invertebrate Biology, Invertebrate Reproduction and Development, Invertebrate Taxonomy, Journal of Experimental Marine Biology and Ecology, Journal of the Marine Biological Association of the United Kingdom, Journal of Natural History, Journal of Plankton Research, Marine Biology, Marine Ecology Progress Series, Memoirs du Museum National d'Histoire Naturelle* (Paris), *Memoirs of the Museum of Victoria, Proceedings of the Biological Society of Washington, Proceedings of the Linnean Society of New South Wales, Proceedings of the Royal Society of London* (series B), *Raffles Bulletin of Zoology, Revista di Biologia Tropical, Sarsia, Smithsonian Contributions to Zoology, Zoologica Scripta, Zoological Journal of the Linnean Society, Zoologischer Anzeiger, Zoosystema.*

## 2. NEWSLETTERS

Included here are some of the more taxonomically or systematically oriented crustacean newsletters of which we are aware. We have purposely avoided listing newsletters that primarily target aspects of crustacean farming, aquaculture, and the aquarium trade.

***Amphipod Newsletter* (see also the Amphipod Homepage)**

Editors as of March 2001: Jim Lowry and Wim Vader

Address: Sydney, Australia (Jim Lowry); Tromso, Norway (Wim Vader)

Homepage: http://web.odu.edu/sci/biology/amphome/

***Anostracan News* (Newsletter of the IUCN/SSC Inland Water Crustacean Specialist Group)**

Editor as of March 2001: Denton Belk

Address: 840 E. Mulberry Avenue, San Antonio, Texas 78212–3194, USA

Homepage: none to our knowledge

***Boletin de la Association Latinoamericana de Carcinologia***

Editor as of March 2001: Guido Pereira (gpereira@strix.ciens.ucv.ve)

Address: Instituto de Zoologia Tropical, Universidad Central de Venezuela, Caracas, Venezuela

Homepage: http://tierradelfuego.org.ar/alca/

***Coral Reef Newsletter***

Editors as of March 2001: C. E. Birkland and L. G. Eldredge

Address: Pacific Science Association, P.O. Box 17801, Honolulu, Hawaii 96817, USA

Homepage: none to our knowledge

***Crayfish News* (Official Newsletter of the International Association of Astacology)**

Editor as of March 2001: Glen Whisson (twhisson@alpha2.curtin.edu.au)

Address: IAA Secretariat, P.O. Box 44650, University of Louisiana at Lafayette, Lafayette, Louisiana 70504, USA (jhuner@usl.edu)

Homepage: http://www.uku.fi/english/organizations/IAA/

***Cumacean Newsletter***

Editors as of March 2001: Daniel Roccatagliata (rocca@bg.fcen.uba.ar), Richard W. Heard, Magdalena Blazewicz, and Ute Mühlenhardt-Siegel

Address: (for Roccatagliata) Departamento de Biologia, Universidad de Buenos Aires, Ciudad Universitaria-Nunex, 1428 Buenos Aires, Argentina

Homepage: http://www.ims.usm.edu/cumacean/index.html

***Cypris* (Newsletter for Ostracodologists)**

(formerly *The Ostracodologist: Newsletter for Ostracod Workers*)

Editor as of March 2001: Elisabeth M. Brouwers

Address: See home page for regional representative

Homepage: http://www.uh.edu/~rmaddock/IRGO/cypris.html

***Ecdysiast* (Official Newsletter of The Crustacean Society)**

Editor as of March 2001: Tim Stebbins (TDS@sdcity.sannet.gov)

Address: City of San Diego Marine Biology Laboratory, 4918 N. Harbor Dr., Suite, 101, San Diego, California 92106, USA

Homepage: http://www.lam.mus.ca.us/~tcs/ecdysiast.htm

***(The) Isopod Newsletter***

Editor as of March 2001: Brian Kensley (kensley.brian@nmnh.si.edu)

Address: Department of Invertebrate Zoology, NHB-163, Smithsonian Institution, Washington, D.C. 20560–0163, USA

Homepage: none to our knowledge

***(The) Lobster Newsletter***

Editor as of March 2001: Mark Butler

Address: Department of Biological Sciences, Old Dominion University, Norfolk, Virginia 23529–0266, USA

Homepage: none

***Monoculus* (Copepod Newsletter)**

Editors as of March 2001: Hans-U. Dahms and H. Kurt Schminke

Address: Fachbereich 7 (Biologie), Universitat Oldenburg, D-26111, Oldenburg, Germany

Homepage: http:www.hrz.uni-oldenburg.de/monoculus

***Plankton Newsletter***

Editors as of March 2001: P. H. Schalk (peter@eti.bio.uva.nl) and S. van der Spoel

Address: P.O. Box 16915, 1001 RK Amsterdam, The Netherlands

Homepage: none to our knowledge

***SCAMIT Newsletter* (Southern California Association of Marine Invertebrate Taxonomists)**

Editor as of March 2001: Don Cadien (dcadien@lacsd.org)

Address: Marine Biology Laboratory, County Sanitation Districts of Los Angeles County, 24501 South Figueroa Street, Carson, California 90745, USA

Homepage: http://www.scamit.org/index.htm

### *(The) Stomatopod Newsletter*

Editors as of March 2001: Tatsuo Hamano (hamanot@fish-u.ac.jp) and Chris Norman (norman@snf.affrc.go.jp)

Address: National Fisheries University, P.O. Box 3, Yoshimi, Japan

Homepage: None

### *(The) Tanaidacea Newsletter*

Editors as of March 2001: Richard W. Heard (richard.heard@usm.htm) and Gary Anderson

Address: Institute of Marine Sciences, The University of Southern Mississippi, P.O. Box 7000, Ocean Springs, Mississippi 39566–7000, USA

Homepage: http://tidepool.st.usm.edu/tanaids/newsletter98.htm

### *Zoea* (Larval development newsletter for carcinologists)

Editors as of March 2001: Klaus Anger, José A. Cuesta, and Pablo J. López-González

Address: Departamento de Ecologia, Facultad de Biologia, Apdo 1095, E-41080 Sevilla, Spain

Homepage: http://members.es.tripod.de/Megalopa/index.htm

## III-B. WEB SITES

Knowing that any such list will become obsolete even before it is published because of the rapid growth of web sites in various areas of invertebrate biodiversity, we nevertheless offer here some of the more useful crustacean-related web sites of which we were aware at the time of printing. Although some of the sites were useful in constructing the current classification, listing below does not necessarily indicate our endorsement nor does it necessarily indicate that the authors of any of these sites are in agreement with the currently proposed classification.

This list is far from exhaustive. It is meant to provide an introduction to the large and ever-growing number of web sites that may be of interest to students of carcinology. Additionally, the list excludes a number of "personal" sites (such as those of Colin MacLay, Jeff Shields, Dieter Walossek, and others), some of which are quite interesting and contain a lot of information on crustaceans as well. A brief selection of these personal sites is given after the alphabetized web page list.

### About Phreatoicidean Isopods in Australia

http://www-personal.usyd.edu.au/~buz/popular.html

A site devoted to these fascinating crustaceans, maintained by George (Buz) Wilson, Australian Museum.

### (The) Amphipod Homepage

http://www.odu.edu/~jrh100f/amphome/

Maintained by Stefan Koenemann at Old Dominion University, Norfolk, Virginia. Nice introductory page leading to the "Amphipod Newsletter," web sites related to amphipods, pictures of amphipods, and various sites about crustacean biology.

### Animal Diversity Web

http://www.oit.itd.umich.edu/bio108/Arthropoda/Crustacea.shtml

This address takes you to the Crustacea pages of the University of Michigan's Animal Diversity web site. Provides general information on several classes, primarily geared to the nonspecialist.

### Animal Evolutionary Pattern Analysis Home Page

http://www.bio.uva.nl/onderzoek/cepa/Default.html#LL

Presents the research activities of a group of scientists allied to the Institute for Systematics and Population Biology, a research institute within the Faculty of Biology of the University of Amsterdam, with links to their ongoing arthropod and crustacean projects.

### Animals4ever

http://www.animals4ever.com/

A searchable and interactive listing, with figures and references, maintained in Belgium, with the goal of eventually grouping "all animals on the web in one place."

### Ant'phipoda, The Antarctic Marine Biodiversity Reference Center Devoted to Amphipod Crustaceans

http://www.naturalsciences.be/amphi/

Managed by the Laboratory of Carcinology at the Royal Belgian Institute of Natural Sciences, with links to the checklist of amphipods of the Southern Ocean, amphipodologists involved with Antarctic fauna, research activities, pictures, and numerous amphipod sites.

### (The) Appalachian Man's Crayfish Photo Gallery

http://webby.cc.denison.edu/~stocker/cfgallery.html

Many color crayfish photographs, plus links to other crayfish sites. Maintained by Whitney Stocker of Gunison University, Ohio, USA.

### Biographical Etymology of Marine Organism Names (BEMON)

http://www.tmbl.gu.se/libdb/taxon/personetymol/index.htm

An interesting site attempting to track the history of taxonomic names of marine species, including crustaceans. Maintained by Hans G. Hansson.

**Biology of Copepods**

http://www.uni-oldenburg.de/zoomorphology/Biology.html#biotable

A page maintained by Thorsten D. Künnemann, with an introduction to the biology of copepods, scanning electron micrographs, copepod systematics, and anatomy of copepods (in preparation).

**Biomedia Home Page**

http://www.gla.ac.uk/Acad/IBLS/DEEB/biomedia/home/home.htm

Very general information on crustaceans (and other taxa) for the nonspecialist.

**BIOSIS—Internet Resource Guide for Zoology (Crustacea)**

(see Zoological Record)

**Biospeleology Home Page: The Biology of Caves, Karst, and Groundwater**

http://www.utexas.edu/depts/tnhc/.www/biospeleology/

Provides information on some cave crustaceans. This site is maintained by the Texas Memorial Museum in Austin.

**(The) Blue Crab Home Page**

http://www.blue-crab.net/

A useful and large resource page, with connections to literature, other sites about blue crabs, and other researchers interested in nearly all aspects of the blue crab, *Callinectes sapidus*. Maintained by Vince Guillory.

**British Marine Life Study Society**

http://cbr.nc.us.mensa.org/homepages/BMLSS

Brief reports of British marine life, with occasional reports of crustaceans and links to other marine life sites.

**Canadian Museum of Nature's Database of Canadian Arthropod (excl. Insects) Systematists**

http://www.nature.ca/english/arthro.htm

A database of Canadian systematists, with scientists organized by area of expertise in arthropods (excluding insects).

**Central Terminal for Crustacean Neuroscience**

http://wwwzoo.kfunigraz.ac.at/crusties.html

A valuable site for everything related to crustacean neurology, with many interesting links to related sites.

**Cercopagis pengoi Page (Cladoceran)**

http://www.ku.lt/nemo/cercopag.htm

A reference page for this cladoceran species; part of the Baltic Research Network on Ecology and Marine Invasions and Introductions, Estonian Marine Institute, Tallinn, Estonia. Contains taxonomic information, diagnosis, line drawings and color photographs, information on population dynamics, and references.

**Cladocera**

http://www.cladocera.uogluelph.ca/

This site, maintained by Paul Hebert, provides a variety of information useful for cladoceran researchers and others interested in the Cladocera. Includes pages on taxonomy, references, researchers, specimen wish lists, tools, and meetings.

**Copepods and Groundwater Biology**

http://www.uni-oldenburg.de/zoomorphology/Groundwater.html

Maintained by the Zoomorphology Section at the University of Oldenburg, this is an overview page with links to Giuseppe Pesce's various groundwater biology sites.

**Crabs Found in Belgium Waters**

http://uc2.unicall.be/RVZ/CrabBook.html

A clever, useful sight for learning about crabs in this part of the world. Click on any crab for further information.

**(The) Crayfish (T. H. Huxley, 1879, 1880)**

http://www.biology.ualberta.ca/palmer.hp/thh/crayfish/htm

T. H. Huxley's classic paper on crayfish in its entirety, including all of the original woodcut illustrations, available online courtesy of Eric Eldred and the University of Alberta, Canada.

**(The) Crayfish Corner**

http://www.mackers.com/crayfish

A lay person site with general information about crayfish, their appearance, behavior, internal anatomy, pictures, and more.

**Crayfish Home Page**

http://bioag.byu.edu/mlbean/CRAYFISH/crayhome.htm

Keith Crandall's website highlighting lab personnel, publications and data, computer programs, lab links, lab tour, extensive crayfish photo gallery, and links to crustacean societies, conservation, and more.

**Crustacea Gopher (U.S. National Museum, Smithsonian)**

gopher://nmnhgoph.si.edu:70/11/.invertebrate/.crustaceans

The gopher menu allows access to "Crayfish," "Isopods," and the "CRUST-L Discussion Group Digests." The "Crayfish" contains 13,000 searchable references. For isopods, see listing under "World List of Marine, Freshwater, and Terrestrial Isopod Crustaceans."

**Crustacea of Lake Biwa**

http://www.hirano-es.otsu.shiga.jp:80/fish-e.html

Images and Japanese names of freshwater crustaceans in Lake Biwa.

**Crustacea Net**

http://www.crustacea.net

Hosted on the Australian Museum website maintained by Jim Lowry, the DELTA (DEscription Language for TAxonomy) taxonomic computer program provides illustrated and interactive keys to identify higher Crustacea taxa, with keys to crustacean families.

**Crustacea Node of the Tree of Life Project**

http://phylogeny.arizona.edu/tree/eukaryotes/animals/arthropoda/crustacea/crustacea.html

This will take you directly to the Crustacea part of David and Wayne Maddison's Tree of Life project. The crustacean section currently is based on Brusca and Brusca (1990).

**Crustacean Disease Information**

http://www.geocities.com/CapeCanaveral/Lab/7490/index.html#crustdis

Part of the Aquaculture Health Page, maintained by Bill Lussier.

**(The) Crustacean Biodiversity Survey**

http://www.nhm.org/cbs/

A site of general interest that includes a searchable, additive, database.

**(The) Crustacean Society**

http://www.vims.edu/tcs

The Crustacean Society Home Page, maintained by Jeff Shields and hosted by the Virginia Institute of Marine Science.

**Crustacean Specimens of the Marine Biological Laboratory**

http://database.mbl.edu/SPECIMENS/phylum.taf?function=search&find=Arthropoda

Crustacean specimens in the collections of the Marine Biological Laboratory, Woods Hole, Massachusetts.

**Crustacés Polynésiens**

http://biomar.free.fr/

Provides a list of species and authorships of Indo-Pacific taxa; many entries are represented with photographs. Maintained by J. Poupin.

**Cryptofauna of Empty Barnacle Shells and Lego Plastic Blocks**

http://www.ex.ac.uk/biology/adrianc.html

Strange but true, an interesting site on an obscure topic, maintained by Adrian Clayton.

**Cumacean Home Page**

http://nature.umesci.maine.edu/cumacea.html

A product of a PEET grant from the U.S. National Science Foundation, this site is maintained by Les Watling and Irv Kornfield (and students) at the University of Maine.

**Directory of Copepodologists**

http://www.univaq.it/~sc_amb/wac.html

Self explanatory; this is a subpage of the Monoculus site.

**Diversity and Geographical Distribution of Pelagic Copepoda**

http://www.obs-banyuls.fr/RAZOULS/WEBCD/accueil.htm

A pelagic copepod site maintained by Claude Razouls and Francis de Bovée at the Observatoire Océanologique de Banyuls, France.

**European Register of Marine Species**

http://www.erms.biol.soton.ac.uk/

A register of marine species in Europe established to facilitate marine biodiversity research and management. Contains checklists of European species, including most of the major groups of crustaceans.

**Ellis and Messina Catalogue of Ostracoda**

http://www.micropress.org

An electronic version of the former looseleaf catalogue from the American Museum of Natural History (Micropaleontology Press). Visitors must go to the catalogues section of the site.

**Epicaridea Page**

http://www.vims.edu/~jeff/isopod.htm#Epicaridea

A thorough page devoted to parasitic isopods, maintained by Jeff Shields, Virginia Institute of Marine Science.

**(The) Expert Center for Taxonomic Identification (ETI)**

http://wwweti.eti.bio.uva.nl/

A nongovernmental organization working with UNESCO and sponsored by the Netherlands Organization for Scientific Research (NWO), the University of Amsterdam, and UNESCO. Includes the World Biodiversity Database (under construction), World Taxonomists Database, and UNESCO-IOC Register of Marine Organisms.

**Fiddler Crabs**

http://www.public.asu.edu/~mrosenb/Uca/

A fiddler crab web site maintained by Mike Rosenberg, with 1,700 references, color photographs, and systematic information, mostly from his recent (2000) dissertation.

**Génétique et Biologie des Populations de Crustacés**

http://labo.univ-poitiers.fr/umr6556/

A research program in genetics and population biology of crustaceans organized through the Université de Poitiers, France.

**Glossary of Morphological Terms**

http://www.nhm.org/lacmnh/departments/research/invertebrates/crustacea/cbs/Glossary_of_Morphological_Terms/index.shtml

A page of the Crustacean Biodiversity Survey, this will eventually be the largest existing glossary of crustacean terminology. Contains multiple definitions put forth by various authors.

**Groundwater Biology**

http://www.geocities.com/~mediaq/fauna.html

Contains many links to groundwater crustacean sites including amphipods, isopods, copepods, remipedes, mysids, spelaeogriphaceans, syncarids, mictaceans, and others. Some links go to specialists' home pages, others contain lists of taxa, still others are in the process of being developed. Maintained by Giuseppe L. Pesce.

**(The) International Association of Meiobenthologists**

http://www.mtsu.edu/~kwalt/meio/

A society representing meiobenthologists in all aquatic disciplines, producing a quarterly newsletter entitled Psammonalia. The site includes several photos of live copepods and links to researchers (including some with expertise in Crustacea).

**International Research Group on Ostracoda**

http://www.uh.edu/~maddock/IRGO/irgohome.html

Includes links to many useful sites of interest to ostracod workers. Maintained by Rosalie Maddocks.

**International Web Site on Terrestrial Isopods**

http://mother.biolan.uni-koeln.de/institute/zoologie/zoo3/terra/homepage.html

This site was still being constructed as of our last check.

**(A) Key to Cladocerans (Crustacea) of British Columbia**

http://www.for.gov.bc.ca/ric/Pubs/Aquatic/crustacea/

Provides keys to the families Holopedidae, Sididae, Daphniidae, Bosminidae, Leptodoridae, and Polyphemidae occurring in British Columbia (approximately 45 species). Published by the Resources Inventory Committee of British Columbia.

**Keys to Marine Invertebrates of the Woods Hole Region**

http://www.mbl.edu/html/BB/KEYS/KEYScontents.html

Chapters 11, 12, and 13 of this series deal with "Lower Crustacea and Cirripedia," "Pericaridan [sic] Crustaceans," and "Decapod and Stomatopod Crustaceans," respectively.

**Laboratory of Aquaculture and Artemia Reference Center**

http://allserv.rug.ac.be/~jdhont/index.htm

The Artemia Reference Center at the University of Ghent, Belgium.

**Large Branchiopod Home Page**

http://mailbox.univie.ac.at/Erich.Eder/UZK/

Eric Eder's site for "everything you ever wanted to know about large branchiopods."

**Leptostraca**

http://www.nhm.org/~peet/

A comprehensive site on leptostracans maintained by Todd Haney (toddhaney@crustacea.net) as part of a PEET project funded by the U.S. National Science Foundation.

**(The) Lurker's Guide to Stomatopods**

http://www.blueboard.com/mantis/welcome.htm

Alan San Juan's stomatopod site at Seton Hall, described by him as "an additional information resource for those people interested in the study and care of stomatopods (mantis shrimps)."

**Marine Crustaceans of Southern Australia**

http://www.mov.vic.gov.au/crust/page1a.html

This excellent guide has been assembled by Gary Poore (Museum of Victoria, Melbourne) as a reference for the identification of a few (about 100) of the numerous species of marine crustaceans known to exist in southern Australia. Richly illustrated with excellent photographs and accompanied by background information on the biology, distinguishing characters, habitat, and distribution of the species illustrated.

**Monoculus—Copepod Newsletter**

http://www.uni-oldenburg.de/monoculus/

The home page of the copepodologist's newsletter, edited by Hans-Uwe Dahms.

**National Center for Biotechnology Information Taxonomy Browser**

http://www3.ncbi.nlm.nih.gov/htbin-post/Taxonomy/wgetorg?id=6681&lvl=10

For locating DNA/RNA sequences of a variety of crustaceans.

**National Shellfisheries Association**

http://www.shellfish.org/

The home page of this association, with links to journals and other activities.

**"Non-Cladoceran" Branchiopod Shrimp of Ohio**

http://www-obs.biosci.ohio-state.edu/f-shrimp.htm

Contains information on anostracans, notostracans, and conchostracans of Ohio. Maintained by Stephen Weeks, University of Akron, Ohio, USA.

**North East Atlantic Taxa**

http://www.tmbl.gu.se/libdb/taxon/taxa.html

Contains PDF files of species checklists, including crustaceans from this region, compiled by the Tjärnö Marine Biological Laboratory, Sweden.

**Orsten and Crustacean Phylogeny**

http://biosys-serv.biologie.uni-ulm.de/sektion/dieter/dieter.html

Dieter Walossek's page introducing the "orsten" fossils (Upper Cambrian of Sweden), Eucrustacea, nonarthropod crustaceans, and more. Includes photographs and drawings of the "orsten" arthropods.

**Ostracod Research Group**

http://users.aber.ac.uk/alm/web/ostrweb2.html

A site maintained by Robin Whatley and Henry Lamb; this is a subgroup of the Micropaleontology Research Group in the Institute of Geography and Earth Sciences at the University of Wales, Aberystwyth.

**Pesce's Home Page/Groundwater Fauna of Italy**

http://www.univaq.it/~sc_amb/pesce.html

Contains information about, and links to, groundwater and speleofaunal crustaceans of Italy, with links to other sites dealing with amphipods, mysids, copepods, and more.

**PHOTOVAULT's Aquatic Crustacean's Page**

http://www.photovault.com/Link/Animals/Aquatic_Crustacia/AARVolume01.html

A commercial site that contains many photographs of various crustaceans.

**SCAMIT Arthropods of Southern California**

http://www.scamit.org/SpeciesList/arthropd.htm

An unannotated list of the species of soft bottom habitats off southern California, maintained by the Southern California Association of Marine Invertebrate Taxonomists (SCAMIT).

**(A) Stereo-Atlas of Ostracod Shells**

http://www.nhm.ac.uk/hosted_sites/bms/saos.htm

A site with information on this and other publications of the British Micropaleontological Association.

**(The) Subterranean Amphipod Database**

http://www.odu.edu/~jrh100f/amphipod/

Maintained by John Holsinger at Old Dominion University, Norfolk, Virginia.

**Systematics of Amphipod Crustaceans (order Amphipoda) in the families Crangonyctidae and Hadziidae**

http://www.odu.edu/~jrh100f/

A U.S. National Science Foundation PEET project maintained by John Holsinger (and his students) at Old Dominion University, Virginia, USA. Includes the Subterranean Amphipod Database.

**Tanaidacea Homepage**

http://tidepool.st.usm.edu/tanaids/index.html

A comprehensive and searchable listing of all

tanaid taxa and the literature in which they were initially described. Maintained by Richard W. Heard and Gary Anderson at the University of Southern Mississippi, USA.

**Urzeitkrebse—Lebende Fossilien!**

http://mailbox.univie.ac.at/Erich.Eder/UZK/index2.html

Contains information on large branchiopods (Anostraca, Conchostraca, and Notostraca) of Austria, maintained by Eric Eder.

**(The) University of South Carolina Meiofaunal Laboratory of Bruce Coull**

http://inlet.geol.sc.edu/~nick/

A meiofauna page, including harpacticoid copepods, maintained by Bruce Coull at the University of South Carolina.

**World List of Marine, Freshwater and Terrestrial Isopod Crustaceans**

http://www.nmnh.si.edu/iz/isopod

Contains more than 9,900 isopod records, all described species of isopods, and a complete bibliography in a searchable Access database. Maintained by Brian Kensley (kensley.brian@nmnh.si.edu) and Marailyn Schotte (schotte.marilyn@nmnh.si.edu) of the USNM, Smithsonian Institution.

**(The) World of Copepoda**

http://www.nmnh.si.edu/iz/copepod/

Contains bibliographic databases for all the literature contained in the Wilson Library on copepods and branchiurans. In total, the website contains four databases: (1) a bibliography of all known copepod and branchiuran literature, (2) a taxonomic list of reported Copepoda and Branchiura genera and species, (3) copepod and branchiuran researchers of the world, and (4) copepod and branchiuran type holdings of the U.S. National Museum of Natural History. Maintained by Chad Walter.

**Zoological Record Taxonomic Hierarchy**

http://www.biosis.org.uk/zrdocs/zoolinfo/grp_crus.htm

The extensive Internet Resource Guide for Zoology provided by Biosis and the Zoological Society of London.

**Zooplankton Sensory Motor Systems**

http://www.pbrc.hawaii.edu/~lucifer/

Contains information on, and links to, research and researchers investigating sensory biology and motor processes and systems in zooplankton of pelagic crustacean and crustacean larvae. Maintained by Dan Hartline and Petra Lenz.

## INDIVIDUAL WORKERS WITH HOME PAGES CONTAINING CRUSTACEAN INFORMATION

**Gary Anderson**

http://tidepool.st.usm.edu/gandrsn/gandrsn.html

A well-designed site with a large number of links to other sites of interest to crustacean workers.

**Raymond Bauer**

http://www.ucs.usl.edu/~rt6933/shrimp/

Highlights his research interests in marine habitats and the biology of caridean and penaeoid shrimp, mating behavior and strategies, hermaphroditism and sex change, antifouling (grooming) behavior, sperm transfer, latitudinal variation in breeding patterns, seagrass fauna, coloration and camouflage, and student research.

**Geoffrey Boxshall**

http://www.nhm.ac.uk/science/zoology/project1/index.html

A single page with general information on copepods, linked to The Natural History Museum, London site.

**Raul Castro R.**

http://members.xoom.com/renrique/copepoda2.html

List of parasitic copepods on Chilean fishes with a list of his publications.

**Paul Hebert**

http://www.uogluelph.ca/~phebert/

Summarizes past and current research and multimedia projects of the lab.

**Wolfgang Janetzky**

http://www.ifas.ufl.edu/~frank/crbrom.htm

Highlights his interests in Crustacea inhabiting bromeliad phytotelmata.

**Gertraud Krapp-Schickel**

http://hydr.umn.edu/g-k/index.html

Highlights her interests in amphipods, plus photos of amphipodologists.

**Colin McLay**

http://www.zool.canterbury.ac.nz/cm.htm

Highlights his research interests in population and marine ecology, reproductive biology, mating strategies, and phylogeny, especially of anomurans and brachyurans.

**Jeffrey Shields**

http://www.vims.edu/~jeff/

The parasitic isopods of Crustacea (Bopyridae, Entoniscidae, and Dajidae).

**Wim Vader**

http//:www.imv.uit.no/ommuseet/enheter/zoo/wim/index.html

A single page highlighting his interests in Crustacea and Amphipoda.

**George (Buz) Wilson**

http://www-personal.usyd.edu.au/~buz/home.html

Research interests emphasizing asellotan and phreatoicidean diversity, with links to many other isopod and crustacean sites.

## III-C. CRUSTACEAN LIST SERVERS

**ALCA-L**

majordomo@fenix.ciens.ucv.ve

List server of the Asociation Latinoamericana de Carcinologia, currently maintained by Guido Pereira (gpereira@strix.ciens.ucv.ve), Instituto de Zoologia Tropical, Universidad Central de Venezuela, Caracas, Venezuela.

**BRINE-L**

http://ag.ansc.purdue.edu/aquanic/infosrcs/brine-l.htm

A brine shrimp (Anostraca) discussion list, maintained by Lamar Jackson and Harold Pritchett at Mercer University, Georgia. Part of AquaNIC, the Aquaculture Network Information Center.

**COPEPODA**

copepoda@sciencenet.com

A list server for discussions of wide ranging copepod research.

**CRUST-L**

http://www.vims.edu/~jeff/crust-l.html

An informal forum for those interested in Crustacea, including their biology, ecology, systematics, taxonomy, physiology, cell biology, culture, etc. Managed by Jeff Shields.

**OSTRACON**

The Ostracoda Discussion List, OSTRACON@LISTSERV.UH.EDU

A list server for discussions of all things ostracode-like.

**The Vernal Pool ListServ**

vernal@sun.simmons.edu

The Vernal Pool Association maintains a list on the EnvironNet server for those interested in vernal pool studies, protection, and education.

## III-D. SOME MUSEUMS WITH CRUSTACEAN HOLDINGS ON-LINE

**California Academy of Sciences**

http://web.calacademy.org/research/izg/

This will take you directly to the CAS Invertebrate Zoology and Geology Department.

**Department of Invertebrate Zoology at the United States National Museum**

http://www.nmnh.si.edu/departments/invert.html

A well-written overview of the history and activities of the staff of the world's largest collection of Crustacea.

**Illinois Natural History Survey Crustacean Biology Information Page**

http://www.inhs.uiuc.edu/cbd/collections/crustacea.html

One of the largest state collections of crustaceans in North America, with a searchable database and a well-designed page.

**Muséum National d'Histoire Naturelle (Paris)**

http://www.mnhn.fr/

Extensive crustacean holdings, but no information available on line yet.

**Natural History Museum of Los Angeles County**

http://www.nhm.org/

The largest natural history museum in the western United States, this impressive institution is also home to the second largest collection of Crustacea in this country. There are an estimated 110,000 to 120,000 lots, containing 3 to 4 million specimens.

**University of California Berkeley Museum of Paleontology**

http://www.ucmp.berkeley.edu

An interesting page that includes mostly paleontological information on arthropods.

**Zoological Museum, University of Copenhagen**

http://www.aki.ku.dk/zmuc/zmuc.htm

A beautiful home page for one of Europe's oldest and most respected natural history museums. The Crustacea collection is extensive and well-curated.

## Addendum

As might be expected in any attempt to be current in a rapidly changing field, several publications or presentations that bear on high-level relationships of the Crustacea have come to light during the final months while we prepared this volume for the printer. In particular, the following presentations dealing with higher crustacean systematics were selected from among the published abstracts of the Fifth International Crustacean Congress in Melbourne, Australia (July 9–13, 2001) (*Fifth International Crustacean Congress—Program and Abstracts, and List of Participants, 2001*): Developmental data in crustacean systematics (Koenemann and Schram); Peracarida (Wilson, Watling, Richter, Jarman, Spears et al., Wilson and Ahyong, Keable and Wilson, Poore and Brandt, Myers and Lowry); malacostracan affinities with insects (K. Wilson); Decapoda (Ahyong and Schram, Porter et al., Brösing and Scholtz, Crandall et al., Richter, Pérez-Losada et al., Boyce et al., Wetzer et al., Ngoc-Ho); Remipedia (Spears and Yager); Leptostraca (Walker-Smith and Poore); Phosphatocopina (Maas and Walossek); Rhizocephala (Glenner and Spears).